GRASS ROOTS AND GREEN TAPE

Principles and Practices of Environmental Stewardship

GRASS ROOTS AND GREEN TAPE

Principles and Practices of Environmental Stewardship

Anna Carr

THE FEDERATION PRESS
2002

Published in Sydney by:
The Federation Press
PO Box 45, Annandale, NSW, 2038
71 John St, Leichhardt, NSW, 2040
Ph (02) 9552 2200 Fax (02) 9552 1681
E-mail: info@federationpress.com.au
Website: http://www.federationpress.com.au

National Library of Australia
Cataloguing-in-Publication entry

Carr, Anna (Anna Jane Lennox)
Grass roots and green tape: principles and practices of environmental stewardship

Bibliography.
Includes index.
ISBN 1 86287 338 0

1. Environmental management - Australia - Citizen participation. 2. Environmental management - Australia - Case studies. 3. Environmental protection - Australia - Citizen participation. 4. Environmentalists - Australia.
I. Title

338.9250994

Front cover photography courtesy of the Department of Land and Water Conservation, NSW.

Typeset by The Federation Press, Leichhardt, NSW.
Printed by Ligare Pty Ltd, Riverwood, NSW.

Contents

To Isabelle, Mietta and our future.
For Charles, Natalie and my past.

Preface

The positive attitudes, unfailing effort and creative spirit of people engaged in environmental stewardship is awe inspiring. This is what motivated me to start researching this topic and is what continues to nourish my curiosity in talking to, with and about environmental stewards. Whether described as activists or volunteers, many seem to embody Ghandi's simple adage, paraphrased here by US activist Sonja Johnson: "We must remember that one determined person can make a significant difference and that a small group of determined people can change the course of history." There is something else too, which brings me to this topic, something best described as a piece of my own history.

When I was eight years old, I sat behind my father driving our old Lanz Bulldog tractor as he cleared the remaining 500 acres of "virgin bush" from our wheat and sheep farm "Allanooka" in Western Australia. It was the late 1960s. While the peace protests, women's and environmental movements were shaking North America, my brother Ben and I watched as the surrounding native vegetation was ripped and bulldozed from our neighbours' properties. Over several years the kangaroos and emus - we used to count them from the window of our school bus - stopped coming to that country. The once rich biological diversity of the sand plain turned to scrub, then to savannah, then into a bleached sandy landscape before it blew away. The road to our farm turned into a beach track as thousands of tons of white "topsand" was deposited there incrementally by the hot easterly winds.

Those memories of rendering our bush with its dainty orchids, egg-and-bacon plants, wattles, and tea-trees into "productive" land still haunt me. When I started talking to others engaged in landcare, watershed restoration or water watch projects, I realised that I was not alone in my sadness at what has happened to land in much of rural Australia. Looking back at what had been accepted agricultural practice is simultaneously uncomfortable and devoid of meaning. Clearing land was commonplace - it was just what happened. Despite this, for those of us who remember, thinking about clearing remains unpleasant. So I chose not to think about it for a long time.

As a teenager, like many in remote parts of Australia, my central aim in life was to get away from the bush, from the village of my primary school years, even from the regional town where I went to high school. I went to the city and counted myself lucky to have left. Staying would have inevitably lead into a trap of marriage to that hot dry land and the landowners, or so I perceived at the time. Not one of my girlhood friends stayed in the area, and most of the boys I knew did not become farmers either. Even if they stayed on the land, they stayed as shearing contractors, truck drivers or mechanics. The land was too marginal, the older generation stayed put, the markets were too distant, the weather too fickle, the commodity prices for wheat, wool, oats, beef or barley too variable. The odds were definitely stacked against young people making an income farming the land on which they had grown up.

My life moved on, to university, into youth issues, politics, regional planning and community and economic development. I was concerned with unemployment, urban decay, and industrialisation. I moved house countless times. Then I moved countries. It wasn't until I was in Canada that I started putting the pieces of the sustainability puzzle together. Meaningful work, lively communities, our "comfortable" lifestyle, all of these depended on the environment. But care of or for the environment alone is an insufficient condition for sustainability. That elusive concept has to be premised upon societal interactions with our natural heritage, whether it is described as a "resource" or a "wilderness". I came home to Australia and set up camp in Canberra. There I was based, but moved around the country conducting the three case studies that are both the research topic for my PhD and the subject of this book.

Having lived in what some consider "an isolated area", (our nearest neighbour in the early days at Allanooka was 12 miles away), issues of scale have always fascinated me. Perspective seems imperative when looking at a particular topic. So rather than concentrate on either the individual, or society as a whole, I focussed on communities of like-minded people working together for environmental stewardship. Situating myself within *rural* landscapes where people were working to find out about aspects of their environment, or act in a remedial or restorative fashion was a logical intersection of my interests and experiences at that point in life. Talking with rural Australians about environmental stewardship was a lot like returning home. In the country kitchens (where the blue and white china and checked table cloths of "country life" magazines were notably absent), the shearing sheds and lounge rooms of those people I was privileged to interview, I knew that I "belonged". That level of connection with these environmental stewards and beyond them to the environment itself has rekindled my trust in restoration.

Having gone away from Allanooka, Walkaway, Geraldton, Perth and then having lived outside Australia for five years, I had picked up an outsider's ability to see life anew in these familiar rural places. But when I returned, it was never as an academic tourist collecting snapshots and data to send back to the city. When meeting environmental stewards (landcarers, water watchers, catchment managers etc), I thought of Jenny, Jan or Jim not as "others" to objectify and describe, but as extended family members of my own rural roots. "They" became "Us". When examining the role that government officers played in these groups, I was reminded of my father's dual role as Agricultural Extension officer working for a government department and local farmer. Government officers were simultaneously a representative of the State, a receptacle of information and advice, and a local person you saw at the pub or the post-office. Hence my tendency has always been to examine these points of interaction. I like to think this book weaves together these diverse perspectives and examines environmental stewardship as a personal challenge, a political goal and a topic for critical inquiry. In writing and researching this book, I feel that I have discovered a way to act against the environmental destruction of Allanooka.

Here, I'd like to explain my use of pseudonyms throughout the book. While describing or explaining events that have occurred throughout the past decade, this book cannot purport to be an historical account of environmental stewardship. Therefore I cannot adopt the tradition of historical scholarship of using people's real names, just the names of the organisations to which they belong. Despite several interviewees' willingness to be identified, I have chosen not to identify any individuals for the sake of consistency and to protect the anonymity of others. However, when using pseudonyms, I have chosen not to use parentheses unless I am describing a fictitious collective. When I say Jenny, or Jan or Jim, I am thinking of a specific individual.

During my "field work", I interviewed 90 people. Two thirds of these were either members of stewardship groups or non-members who lived nearby, most of whom supported the stewardship ethic. For these people, the environmental movement is not something with which they could readily identify. For them, it is less of a movement, with all its political implications attached, and more of a transition, a step toward sustainability. Environmental stewardship, looking after their land, their waterways and watersheds, was a step beyond "their own backyards" but not so much of a leap into the unknown world of grassroots activism and advocacy that they felt uncomfortable. To be a "greenie" in rural Australia is still an anathema to most. The remaining third were government officers, facilitators, bureaucrats, administrators or funders. Their perspective is equally important in sustaining the on-ground efforts. Occasionally I report government efforts as less than advantageous to local stewardship groups. Where points of disjuncture or exigency are reached in the text, I mean them less as implied criticisms than as invitations to review government policy or practice ... necessitating the use of green tape and not red.

Acknowledgements

I never really understood why some authors made such a fuss of thanking their editors... until now. Chris Holt, Margaret Farmer and Clare Moss of the Federation Press have been an unending source of solidarity and resourcefulness. Through the births of both my children, household relocations across the country and back again, the winning and losing of grants, through job security and insecurity, Federation Press has been there. For their unerring support, I thank them sincerely.

Without the financial contribution of grant monies, stipends and salaries, books like these don't get written. I acknowledge financial support provided by the Rural Industries Research & Development Corporation, the Rockefeller Foundation and the Australian National University. The Australian Research Council provided a secondary, although important, means of support.

The Centre for Resource and Environmental Studies (CRES), within the Australian National University is a special place. In addition to providing the financial, material and intellectual space and resources in which to think and write, it accommodates those people who make life as an author into an academic adventure. Henry Nix, Helen Ross, Libby Robin and Steve Dovers have thoughtfully provided intellectual rigour and academic enthusiasm on various occasions. I also acknowledge McComas Taylor for publication information with panache and Susan Kelo, for giving practical tips about word processing. Ex-CRESites Vicki Strickland and Ettie Oakman are a formidable team (both separately and especially together!) and deserve particular acknowledgement for the meticulous and systematic formatting and reformatting of each chapter. For inspirational talks on the nature of nature, I was lucky to have had the chance for tea with Nugget Coombs and still see Val Brown and Stephen Boyden from time to time. Optimism in the face of environmental adversity could always be extracted from my colleagues and occasional mentors Deborah Bird Rose, John Handmer and Ian White.

When this book was still (only) a PhD thesis, comments made by examiners Sally Lerner, John Harris and Brian Roberts proved both insightful and encouraging upon its transition to a more accessible text. Even earlier than that, I owe much to the careful and critical appraisal of the near-final educational product by Richard Baker of the School of Resources, Environment and Society. More recently I am grateful to the more experienced critique of Libby Robin of CRES and one other anonymous reviewer.

Other people in or around the university who have assisted this project include Roger Green of Greenwords for sub-editorial polishing. Robin Connor, Jason Evans, Steve Leahy, Karl Nissen and Phil Greaves deserve medals for computing wizardry of various denominations. I am grateful to Val Lyons for producing Figures 3.1 and 3.2. Other neighbouring academic colleagues with whom I have had many fruitful and interesting interactions include Ken Johnson, David Dumaresque and Ian Hughes.

Past and present students, activists and environmental professionals have been a constant source of stimulation and motivation. Whilst I am unsure of my abilities to remember all who lent ears, kind words or moments of jocularity, I am sure of the debts I owe to my friends and colleagues along the way. These include Alice Roughley, Jamie Allnut, Deborah O'Connell, Nick Abel, Ruth Beilin, Sarah Ewing, Helen Neave, Andrew Campbell, Helen Alexander, Fayen d'Evie, Paul Murfitt, Andrew Taplin, Kate Ord, Jennifer Andrew, Tara McGee, Sureeporn Punpuing, Anuchat Poungsomlee, Jacquie Tracy, Nora Galway, Jenny Rush, Chris Nobbs, Rae Fry, Anne-Marie Wilson, Steven Bygraves, Meg Keen, Irene Guijt, Donna Hazell, Peter Hazell, Lisette Van Vliet, Su Wild River, Jenny Kesteven, Judy Clark, Jane Mogina, Geraldine Cusack, John Mulquiney, Rosemary James, John Lumb and Sarah May. Both on and off campus, I have enjoyed the special friendship of Catherine Mobbs, for being a workmate as well as a playmate. Barbara and David Vernon played an important supportive role at various stages in the early, more contemplative, stages of this thesis-become-book. Others I am pleased to count as friends include Tamsin Kerr and Ross Annels for reminding me that juggling family and work responsibilities was always going to be a challenge and for many hours of their company while "it" was in progress.

I would like to recognise the special friendship of my writers group, especially Jennifer Curtin, Nicki Mazur and Gerlese Akerlind for the time we have spent together and the supportive writing context we have engendered. Many thanks are due to my co-authors on two other writing projects, Roger Wilkinson and

Laura Stocker for understanding the delays brought about by my production of this book and two babies in not-so-quick-succession.

Outside the bounds of Canberra and the university system, many others have contributed to this work in numerous ways. In Western Australia, I acknowledge the role played by Geoff Syme for allowing and promoting time donated to authorship when there was little to give. In South Fremantle, I am eternally grateful to Ben Carr and Chrissie Lauritz for giving me space to stay in their house by the sea. The welcome and care they extended to me when coming "home" to family was almost always warm and inviting.

An important aspect of the research has been the opportunity to work with people who share a commitment to the ideals of community development, land ethics, grass-roots activism and environmental stewardship. During my visits to stewardship groups I made many friends and renewed my love of rural Australia. I am especially grateful to Jan Star, Keith Bradby and Ronniet Orlando of Serpentine Jarrahdale. In Downside, near Wagga Wagga, Cheryl Paton, Tony Dunn, Barbara Carmichael, Max and Helen Chamberlain are all wonderful people whom I thank very much. Lastly, On Cape York, I am grateful to Bob and Katie Pond, Viv Sinnamon, Ben Westerberg, Bruce McCarthy, Hilary Kuhn and Sarah Connor of the Mitchell River Watershed group. These people not only contributed their time and hospitality, they continue to influence me with their love of local places.

Lastly and most importantly, I would like to thank Glenn Johnstone for countless acts of consideration, organisation, clarification, domestication and especially for his uplifting and unflappable spirit.

1 Introduction

> When I was younger – I always thought that these people in government departments could take care of it [the environment], but now I'm seeing that sometimes they get off the track ... We thought they would have had things more under control than that. I'm starting to realise that they're not much smarter than us – even with training, they still make mistakes. (Salim, a Mitchell River Watershed group member)

Like the proverbial lead balloon it dropped noiselessly but loud enough for everyone to know it was a real clanger. There was a sudden lull in the conversation. Moments before, the council board room (complete with a portrait of the Queen and an oval mahogany table) had been a hive of activity; plans, dates, "action-items". Other members of the stewardship group now turned towards the speaker looking slightly stunned.

Yet there was no need to question him further, everyone knew it to be true. Leaving environmental stewardship solely in the hands of government officers was not a smart move.

The small group of hard core Water Watch members had been meeting now for about two years. Their quest had been to prove to the State government that phosphates from their soils were not responsible for leaching out of the paddocks into the creeks and drains and subsequently into the Serpentine river. A few months earlier it had turned the estuary at Mandurah into a bright green algal bloom. What may have started off as a reaction to government as Big Brother (the algae blooms were not our fault because it was not phosphate run-off from farms in the headwaters of the Serpentine Jarrahdale catchment that were causing the problem) soon became an invitation to Uncle Sam, Tom Cobbly and all to help find solutions to a much deeper problem ...

Toward their original aims, the group had successfully completed two seasons of water quality monitoring. Their strategy had been to test water quality at regular intervals along the maze of creeks, streams and irrigation channels back up the catchment to trace the source of phosphate emissions. Now they realised that having captivated community concern and – having finished their local detective work – they had a new and bigger project. That was how does a community group interest government authorities in "the big picture": environmental sustainability of their whole area.

This book is about bottom-up environmental management. It is written for those at the grass-roots – Kerry, Jenny, Louise, Robbo of Water Watchers plus the members of the Downside Landcare and Mitchell River Watershed community groups who were my "guides" in this quest. Their authority was steeped in local knowledge, in their emotional and physical ties with the landscape. They assumed responsibility for environmental stewardship. It is also written for those government employees who want to facilitate local community action on environmental initiatives and who have clearly reached across the grass-roots and green-tape divide into the middle ground.

The rural crisis

Rural community groups all over the world are responding with determination and creativity to local environmental crises. These crises are evident in the social, economic and ecological fabric of the North American prairies and in the Australian outback, from the African savannah to the highlands of Thailand. Ecologically, many parts of the world are experiencing decreasing soil fertility, water quality and biological diversity, while at the same time experiencing increasing acidity, salinity and invertebrate pest populations. In North America, the ecological alarm button was pressed in the late 1960s concerning the health of the Great Lakes system. Nearly 40 years on, the Great Lakes are measurably healthier and thousands of smaller lakes across New England are being monitored by both individuals and the concerted efforts of community groups working together with government agencies.

Economically, average Australian broadacre farm cash income and the total number of farmers have both dropped over the last decade while the average age of remaining farmers has increased. The vice president of the Institute for Local Self-Reliance in the United States recently stated that one thousand farms have gone out of business every week in that country since 1950.[1] I have heard one farmer speak at an Australian Landcare conference about not being able to replace shoes for her children. Reminiscent of the 1930s mid-west United States drought many farmers have been forced to walk off their properties by banks in their haste to foreclose on loans. In Australia, low returns to producers, rising input costs, growing farm indebtedness and falling land values have all contributed to a situation characterised by increasing stress, fear and uncertainty, decreasing trust and diminishing social networks. Smaller communities which have traditionally been held together by the "pub", the school, the church, a sports club and a shop are undergoing "structural adjustment" of local economies, a euphemism for economic decline. Over many years the slow demise of many of these cornerstones of rural Australian society has seen social services relocated to bigger centres or abandoned completely. Coupled with the ease of access to urban centres, much of rural social life has been lost, leaving some farming communities socially isolated and peripheral to towns and "townies".

In the past 25 years, technological, economic and social change has hastened the demise of these smaller communities. While in the post-war era "closer settlement" policies were responsible for higher rural populations, now rural services have been recalled. Many schools have closed and pupils bussed to larger centres. Agricultural extension officers can no longer afford the luxury of one-to-one consultations and have become confined to offices, becoming known to farmers as the "bloke on the other end of the phone". A combination of farm amalgamation, international trends toward large-scale agribusiness and technological change has transformed the farm sector and the rural landscape.

Socially, there is teenage suicide, alcohol abuse, domestic violence, unemployment, lack of access to adequate health and education services, language barriers, inequality in gender relations, and lack of child care facilities. This is true not only in Australia but in many parts of the rural hinterland in the "developed" nations of Canada, New Zealand and the United States. All these factors contribute to the global rural crisis and many have come about as a direct result of economic and environmental factors.

Enter the community group

The causes of these social, environmental and economic problems are complex and interrelated. One of the "solutions" currently practised internationally is the widespread establishment of community groups. In developing countries many rural community groups are initially established by international environmental non-government organisations (ENGOs). In developed nations community groups may be facilitated by government or national advocacy organisations. More frequently they self-organise in response to a local threat. It seems there is a group for all occasions. Footrot groups have formed to control that disease in local areas. Computer aided livestock marketing groups have formed to market their stock. Producer cooperatives are now a recognised entity in most rural communities, and it does not stop with economic incentives. Socially, rural communities have always enjoyed their sports clubs. Hall committees are a feature of almost every small place privileged enough to own a community hall. Interestingly, progress associations seem to have given way to heritage committees - often with many of the same members. And now there is another incentive to form a group in rural Australia - the environmental imperative. Landcare groups, catchment management groups and water quality monitoring groups are proliferating all over the country. Tree-planting groups are no longer seen as "greenies turf" and dune care or coast care groups are commonly valued as worthy stewardship enterprises alongside many others.

Since 1990, numbers of landcare groups alone have grown steadily to around 4500 groups as we begin the new millennium. The percentage of farms where a family representative is a member of a landcare group has also risen which indicates the growing interest in such groups and the success of environmental publicity campaigns. This trend is not peculiar to Australia, although it may be more pronounced there. Community-based environmental restoration projects have taken place all over the world. Kenya's Greenbelt movement, the Philippine's Bamboo project and Canada's Karner Blue Butterfly program in Ontario are all good examples. Consider, for example, the range of North American and Australian environmental interests illustrated in Box 1.1 by a handful of project titles chosen randomly from web-sites and other references to stewardship groups.

Box 1.1 Diversity of stewardship groups

- Friends of the Koalas Inc. (AUS)
- The Steelhead Society Habitat Restoration Corporation (CAN)
- Warroo Landcare Group (AUS)
- Greenlink Box Hill (AUS)
- Friends of Rose Creek (CAN)
- Virginia Save Our Streams (USA)
- Jordan River Stewardship Project (USA)
- Revegetation in Candlebark Park (AUS)
- Mamadawerre Outstation Dust Suppression Project (AUS)
- Glossy Black Cockatoo Habitat Expansion Project (AUS)
- School Yard Habitat Restoration Program (USA)
- Missouri Stream Team (USA)

In the 1990s, there was an active shift on the part of governments to devolve responsibility for economic development, environmental management and most social services to rural communities. Some argue that they are also attempting to absolve themselves of these responsibilities - giving responsibility without power.[2] Economic choice drives more and more individual farmers' decisions in an increasingly deregulated rural economy. For example, with the closure of the Australian Wheat Board, grain marketing has become the individual's responsibility. Deregulation has also taken place in the Australian beef and sugar industries with widespread dairy deregulation occurring in the next wave of new public managerialism. Government funded agricultural advice is now seen as too expensive to be delivered on a one to one basis. Now, agricultural bureaux (groups of farmers) receive more pre-packaged information at field days and seminars than ever before.

As the government has got out, community groups have filled the vacuum and now play a vital role in sustainable environmental management. Community-based environmental groups are flourishing as the bottom-up approach is recognised as an important tool for working toward sustainability. They are becoming increasingly involved in environmental management and are more assertive in decisions about local environments. For this reason they will no longer accept "holier than thou" arguments from traditional authorities such as government departments. Stewardship group members all over the world like Kerry of Water Watchers and Salim of the Mitchell River group are recognising that governments, at all levels, do not always know best, or even have a vision of sustainability. *Caring for the Earth,* the World Conservation Strategy released in 1991, identifies community-based environmental management as one of the key principles towards building a sustainable society. However, with the meteoric rise in the rate of acceptance of "sustainable development" (as a concept - if not a practice), increasing and perhaps undue pressure has been placed on community groups to assume total responsibility for their local environments.

With less government funding available to groups and more groups becoming established, the private sector has become more involved in environmental management. For example, the Fertiliser Industry Federation of Australia has sponsored landcare through endorsements in the *Australian Journal of Soil and Water Conservation.* Advertisements for fencing equipment companies regularly appear in landcare conference proceedings. The National Farmers' Federation (the peak Australian organisation representing many farming industries) is also heavily involved in the quest for sustainable agriculture in Australia through the promotion of landcare and other community initiatives. Australian private sector initiatives, however, pale in comparison with the extent of corporate and philanthropic interest in North American environments. Public-private partnerships are very much a focus of North American efforts in environmental stewardship. For example, the model forest concept from Canada emphasises local job creation in cooperation with private industry. The latest model forest in Mexico provides a winter home for the monarch butterfly; its twin goals are sustainable forest use and habitat protection.

When equipped with skills, resources and opportunities from their own communities as well as external support, local groups are empowered to combat the rural crisis. Gradual change toward sustainable environmental management can occur. Community group action is very important because it takes into consideration the importance of local people's roles in attempting to slow the rate of environmental degradation by increasing the rate of change in local strategies for environmental problems. It also takes into consideration the relationship between knowledge, attitudes and behaviour in determining new goals for sustainable management of natural resources. Community groups operate locally and incorporate local values, attitudes and world-views in designing action strategies. Kerry, Salim, Louise, Robbo and Jenny, like other environmental stewards, are integrally linked to the spatial and temporal context in which they act – thereby using their local knowledge of community residents' skills to undertake environmental restoration. Since 1990, the growing number of conferences on environmental stewardship generally and the relationship between restoration ecologists at agency and community level specifically attests to the importance of this field. For example, in the United States there are regular National Volunteer Monitoring conferences and there have been recent conferences in both Canada and the United States on stewardship and conservation.[3]

However, despite the emergence of large numbers of community groups undertaking environmental projects, there are two fundamental problems for communities involved in environmental management. First, significantly more attention should be paid to the nature of the group process itself, how a group forms, what keeps it motivated, what features of group dynamics contribute to on-ground change and what hinders the group process. For anyone who wants to join a stewardship group, it is important to understand *how* groups work in order to enact environmental stewardship in practice. Second, there should be more attention focused on stewardship groups' relations with government. In this book I briefly examine the group process issues, then focus on the point of contact between stewardship groups and government agencies in an effort to enhance the relationship between bottom-up and top-down management styles.

Of critical concern to me is the extent to which stewardship groups are captured by the state and used to implement sustainable development policies at minimal cost – under the banner of increasing community participation. Alternatively, it could be argued that they are self-directing, autonomous organisations upon whom society must depend in the future if we really want to work toward sustainability. Closer attention must be paid to the principles and conditions under which groups can actively contribute to sustainable management of local environments. In the face of so many groups starting up, all confronting large numbers and types of environmental management problems, and all involving a variety of stakeholders and types of resources, there exists a significant opportunity here to make a contribution to policy development in this area.

Public awareness of the importance of the environment may be responsible for increased stewardship action in rural communities. More community participation in the environmental arena may also be due to a renewed sense of urgency in the need to act to protect our environments, fragile and complex as they are. Perhaps stewardship groups are attracting membership with the growing value shift in modern society that mourns the loss of a sense of community with the push for rationality and efficiency. Whatever the cause, it is clear that there is more international attention to participatory democracy in environmental management than ever before.

There are many questions that still require answers here. Is community involvement in ecological restoration being touted as the cure-all or panacea of sustainability? Is government really handing over resources to the local level, or is it abdicating its environmental responsibilities? What other forms of government support are needed to "enable communities to care for their own environments'?[4] These questions cannot be answered until the merits of community involvement are considered more critically by policy makers. Despite all the rhetoric about landcare and other environmental community-based organisations caring for the land and creating environmental awareness, what do governments really understand about group process? How do rural environmental groups work? What type of changes are they likely to make to the landscape and how will these impact on local people? It seems that environmental stewardship groups have become institutionalised without any apparent concern for how or why they work, as long as they are "fixing the problem".

In addressing these questions, my specific focus is on the experience of three Australian stewardship groups (sometimes referred to as community based environmental management groups) who worked predominantly in rural areas. Members of these groups offered me a rich source of stewardship experience, ideas and inspiration. However, wherever possible I also refer to international examples[5] from developed nations and examine trends, comparisons and original cases of stewardship from around the world.

The three community groups described in this book reflect a combination of different environmental issues, places, people and management styles. Water Watchers in Serpentine Jarrahdale, Western Australia, monitored water quality to determine phosphorus loading. The Downside Landcare group in New South Wales is primarily concerned with dryland salinity. Lastly, the Mitchell River Watershed Management group in far north Queensland is concerned with the conflicting management interests and environmental problems of an entire catchment reaching over 75,000 square kilometres (see Box 1.3).

Box 1.3 Some of the environmental problems in the Mitchell River Watershed perceived by group members

Biophysical problems	
Weeds	rubbervine chiney apple
Mining	stream-bank erosion environmental disturbance heavy metal leaching
Ecological	bushfires loss of species
Agricultural	salinity issues water quality issues fragile soils
Socio-economic and cultural problems	
Economic	over-grazing drought policies
Government	lack of coordination conflicting policies lack of data lack of knowledge
Aboriginal	poor communication lack of cultural respect lack of community awareness
Tourism	over-fishing litter problem
Urban development	urban sprawl town sewerage

The problems are diverse. Some communities have many environmental concerns, others have fewer, but perhaps more serious ones. This is compounded by different subsections of the community viewing environmental degradation in different ways. Sometimes a community is unified (dairy farmers for example) and the problems are diverse or run "off-site" across boundaries. Sometimes the problems are distinctive and localised, but the community is diverse and divided. Often an agent of change (for example a government project officer) is brought in to "fix" something and may then be criticised for bringing about changes. New policies may be enacted which address current issues, but some local authorities are too afraid to implement them in case they are not appropriate in five years' time. Fear, ignorance and uncertainty definitely play a part in local environmental management. This is especially true when there are many different factions entrenched in decision-making "for the common good", and that value-driven construct is far from being clear.

Traditionally the environment movement has attracted individuals (activists and "greenies") *or* large groups operating at State, national or international planning levels (eg Greenpeace, The Wilderness Society or the Sierra Club). Now it is timely to learn about environmental stewardship from smaller groups within the community. This is not to deny that critical investigation of environmental policy operating at the macro level is imperative. Yet, while I am aware of the arguments for institutional reform on a broader scale, here my particular concern is with documenting and appreciating local people's environmental actions.

A word about structure

Key elements and ideas behind community, environment and sustainable development are introduced in Chapter 2. This chapter is deliberately eclectic and discusses these terms such that we can develop a common language. The concept of stewardship has a history. If we understand that history and share a language, we may better understand stewardship groups in practice.

For stewardship groups to contribute anything to sustainable environmental management, it is important to understand why groups start and who belongs to such groups. A *profile* of group membership in Chapter 3 establishes who they are. Likewise, it is important to learn from the way members interact with each other. The *practice* of community-based stewardship is outlined in Chapter 4. The answers to these first questions may then help us to understand the underlying *principles* and characteristics of stewardship groups. Instead of writing about each group in separate chapters, all three groups are covered simultaneously. In this manner, comparisons may be made about motivations for joining groups, decision-making structures, conflict resolution techniques and other aspects of group practice. Chapter 5 examines governments' role in local environmental stewardship. It looks at both the broad policy context and the specific circumstances of government officers' contact with stewardship groups.

The principles of sustainable environmental management in each community group are discussed in Chapter 6, based on different aspects of group function examined earlier. Chapter 7 presents an overview of community-government interactions within an environmental arena. It illustrates the strengths and weaknesses of both bottom-up and top-down approaches and poses some alternative "middle-ground" options for collaboration. It is not intended to be prescriptive. It is, however, intended to be used as a tool to describe and analyse community groups.

Underpinning each chapter is this recurrent theme: neither community groups nor government agencies can undertake sustainable environmental stewardship alone. Both bottom-up (community) and top-down (government) approaches are necessary to work toward the middle ground and combine the resources, skills and knowledge of each so that opportunities for action can emerge.

Endnotes

1 David Morris (1999) "Small is still beautiful'. *Utne Reader* No. 93 May-June 1999: 24-27.

2 See Lawrence 2000, Lockie 1993, Vanclay 1994, Martin and Woodhill 1995 for a good overview of the complex balance between government roles and community rights in rural policy.

3 LAKELAND Stewardship Conference hosted by the Federation of Ontario Cottagers Associations (*www.foca.on.ca/Infobase/Environment/Stewardship.htm*); Caring for our land: stewardship and conservation in Canada (*www.uoguelph.ca/~claws/conference*); Steps toward stewardship: Ponderosa Pine Ecosystem Restoration and Conservation (*www.for.nau.edu/ecorest/conf/present.html*).

4 IUCN/UNEP/WWF 1991: 11.

5 These were drawn from developed nations in order to aid comparison. Developing countries experience of stewardship is different because of their work with international aid agencies and non-government organisations.

2 Towards a common language

> I sometimes feel that I don't always know what they're talking about, but that doesn't stop me from going ... They're a nice group of people, pleasant people to deal with ... (Daphne, a Downside Landcare group member)

The importance of speaking a common language, an ordinary everyday sort of language cannot be overemphasised in stewardship practice. All too frequently our different specialisations come to the fore. Farmers talk of rotational grazing, surface run-off, water tables, price squeeze and phalaris. Agricultural researchers speak of pasture mass, total P, storm events, infiltration capacity, remnant "veg" and evapotranspiration. Academics, well they talk a language unto themselves according to many environmental practitioners, and that language changes from discipline to discipline. The point is we may well be unintelligible to others if we confine ourselves to only talking with people like ourselves. But in matters environmental, surely we need more people with different ideas, different backgrounds, different experiences to offer their skills and unique perspectives to stewardship practice? The environment is not a singular entity nor a collection of biophysical phenomena, it can be understood and experienced in a variety of ways. Most importantly, if we are to communicate concern about the environment and its stewardship, we need to be able to talk to each other in simple language, even though we may wish to convey complex ideas.

To better appreciate both the language and practice of environmental stewardship, there are four key ideas which help explain the context in which this new and rapidly evolving field is emerging; sustainable development, community, environment and integration. *Sustainable development* or sustainable management of our environments has set the context for debate about environmental stewardship. The notion of *community*, how it is defined and its links with the fields of community development and public involvement provides key insights into the way in which grass-roots environmental groups are working. Discussion about the *environment* provides both a back-drop as well as a focal point against which the energy of many environmental stewardship groups is spent. Without some *integration* of these ideas, community-based environmental management loses an holistic perspective and makes little sense to those outside this broader context.

Sustainable development

Sustainable development[1] has become a global catchcry, at least since *Our Common Future* was released by the World Commission on Environment and Development in 1987. Yet it still has very little real meaning in everyday use, despite extensive government expenditure, large numbers of academic papers and unlimited discussion and consultation on the subject.

At the grassroots of society however, sustainable development has enjoyed a far longer history – it just wasn't labelled as such at the time. In Australia community groups such as the River Murray League have been "doing something" about various environmental and other local issues since the 1890s. Landholders in New South Wales in the early 1900s, for example, formed pasture protection boards to jointly undertake weed and rabbit control. River improvement trusts were established in the 1940s and 1950s and some are still operating, albeit in different forms. Many groups in recent times have gone through a long, expensive and demanding community consultation process in the name of catchment planning or developing salt action strategies. This longer history of grass-roots action has occurred on a small scale in most industrialised nations. In the last quarter century though it has speeded up and scaled out with globalisation, market-deregulation, the Internet and the growing population of ecological citizens. Internationally, community groups self-organising to "fix-up" their local environments through the promotion of participatory natural resource management.

Australia too has been subject to the trend of handing responsibility for ecologically sustainable development (ESD) over to "the community". Since 1990 there have been many attempts by State governments to involve communities in environmental projects such as Salt-Watch, Water-Watch and Landcare projects. The critical point here is that there is a big difference between the State abdicating its environmental responsibilities to community groups and communities (who want to take local responsibility for their environments) being supported by the State. Governments ask community groups "how would you like us to tell you what to do?" more frequently than they ask "how do you want us to listen to you"? Despite most groups believing themselves to be subject to the latter approach, without the concomitant devolution of power and resources, many groups end up providing "volunteer" support to State agencies who are running their own agendas.

At the same time, political systems of democratic representation are not always as flexible and committed to specific sustainable development strategies as individuals and groups working "on the ground". It is vital that governments set policy directions in light of international trends and create visions for a sustainable future. At the grass-roots however, it is frequently the case that *adequately resourced* vigorous, well informed and forward thinking communities are able to implement ESD strategies more quickly and with more attention to the specific environmental problems of their area than the State.

Bypassing the expanding array of definitions of sustainable development and philosophical arguments about the difference between sustainable development and sustainable management, there are four themes which are essential for the practice of sustainability – both locally and at national and global scales:

- A *change* in environmental perceptions and action is vital. The time has come to sever connections with business-as-usual, environmental sustainability must be brought about via change.
- People must be involved in making the necessary changes come about, not simply through governmental approaches, business approaches or any one single approach, but through an *integrated* or partnership approach to sustainability which recognises the role played by community groups and voluntary organisations.
- Institutional structures that meet the needs of landholders and wider society are being based in the community where local knowledge and local history can play an important part in decision-making. However, it is important to recognise that neither top-down nor bottom-up approaches will work in isolation, *multilateral* and cooperative efforts must be made.

- Many of the current problems derive from a lack of basic understanding of the environmental resource issues involved. Therefore *holistic* approaches are necessary which start with the problem as perceived by those experiencing it. To promote cooperation between and among volunteers, different State agencies, Federal Government, environmental scientists and others it is vital to create an atmosphere of mutual respect and tolerance.

Stewardship

The stewardship approach to sustainable environmental management embodies many of these themes and recognises that humanity is but one element of a complex web in ecology – not centre stage. Sally Lerner in her Canadian book *Environmental stewardship: studies in active earthkeeping* draws attention to the root of the word. She states that good stewardship implies "caring for, maintaining well-being, being vigilant, accepting responsibility [and] understanding the importance of accountability".[2] Given these qualities it is closer to defining the environmental practice of many local groups than government-sponsored environmental management programs will ever be.

Stewardship also implies a relationship with the earth that is based on respect for nature, and a current and ongoing commitment to "active earthkeeping" akin to a custodial or guardianship role. For some, stewardship implies a moral or religious responsibility to life on earth, as part of nature, not having dominion over nature.[3] For others, it is not only about identification and connection with our little blue-green planet, but also about "shared lifestyle preferences and beliefs about the way in which common property resources contribute to a unique quality of life for the residents"; and "beyond an individual connection to encompass collective responsibility".[4] Local environmental stewardship is critical to sustainability. Volunteers regularly spend a great deal of time in stewardship activities such as monitoring flora and fauna, writing press releases, raising funds, planting trees and revegetating river banks amid a wide range of other tasks. Without the input of people like Daphne, much would be lost in the cause to address environmental problems. So it is imperative that Daphne's contribution is recognised, her skills are valued and every effort made to include her in stewardship practices. Besides a common language, a sense of community is one of the things that makes stewardship attractive for Daphne.

Community

Environmental stewardship groups who have a tangible sense of community take on a wider variety of tasks and complete them more often than their counterparts who are bereft of this solidarity. *Community* is one of those useful all-purpose terms which, like brand names of common household items, have become ingrained in common parlance. In the 1950s, the term *community* had been defined in no less than 94 different research settings.[5] With the advent of globalisation and the Internet there must be many more such definitions by now. However, clearly unlike Kleenex or Hoover, *community* has no fixed meaning, but many "desirable" characteristics: non-violence, direct democracy, self-reliance, local culture, appropriate technology and equality to name a few.

Definitions of community are commonly based on place-centred or people-centred ideas. Perhaps the easiest to visualise is the notion of community as a geographic place; for example Wagga Wagga, Australia or Medicine Hat, Canada. Community defined in terms of people is usually thought of with reference to "like groups", for example the gay community or the international community of physicians. It may also be thought of as the people who both live or work in a specific place and share common ideals such as some religious or intentional communities. One of the more typical uses of the term community connotes the feeling of connectedness or togetherness, something which unites people in a common cause. Yet that "perfect unity" is surely no more than an idealistic pipedream when one considers the underlying ambiguities and conflicts present in the use of the term in the media for example.

A key point here is that one person may belong to many communities simultaneously, which serves to increase the underlying complexities of this concept. Ambiguities and nuances of interpretation raised, for example, by someone who is gay or lesbian, shoots ducks and is a member of the clergy in "the rural community" are especially important to make explicit. He or she may be a member of a gay and lesbian choir whilst also subscribing to *Duck Shooters Anonymous*. In other words, assumptions about what makes up "the community" are always subject to a diverse array of political and sexual interpretations (among others). Thus when concepts of community are not adequately defined and subsequently become administrative units in the eyes of government, there is more confusion. Some of this confusion arising over use of the term "community" is clearly illustrated in Australia with reference to the Murray-Darling Basin Commission's Natural Resource Management Strategy. The Strategy officially recognises "communities of common concern" as legitimate local groups working toward sustainability. However, it would be a categorical mistake to assume that *community* and *community group* simultaneously represent both a social organisation and the whole locality. As one irate member of a salt-action group wrote:

> We are heavily involved in Landcare and similar rosy-glow movements in creating Community awareness – but let's really let the Community know that it has been riding very heavily on the farmer's back over the last 30 years.[6]

Obviously some people believe that if there is a community, then it is somewhere "out there", having little to do with the daily reality of farming life, for example.

The concept of what characterises *community* is a value laden one since it is determined both culturally and socially and depends as much upon what proponents of the concept think it should be as upon what observers think it is. It evokes "warm and fuzzy" feelings among community developers, planners and citizens alike. Indeed, community work has been heralded as democratic, participatory and inclusive and therefore able to increase social capital in rural regions. Since the historic coalition in Australia on land degradation issues, between the National Farmers' Federation (NFF) and the Australian Conservation Foundation (ACF), landcare has been similarly championed as the ultimate model of community cohesion and environmental management. However there are good reasons for challenging this assumption and exposing the facade of these "community" organisations whose superficial cohesion may cover up profound power imbalances between the rural elite and rural poor. Sometimes the only unity and cohesion these types of landcare groups display is that necessary to capitalise on the availability of additional government funding for local community projects.

There is widespread confusion and mistrust about the use of terms like sense of community, community participation and community involvement. There is also a dark side to *community*. Contrary to the perception of "sense of community" as "maudlin togetherness, a tear-soaked emotional drippiness that misguided do-gooders seek to experience",[7] there is evidence that a highly developed sense of community may give rise to exclusion, elitism, polarisation and even cliques and cults. Community psychologists refer to recent gains in popularity for such groups in the United States, such as the Klu Klux Klan and security conscious gated housing estates. In Australia, the mild mannered Ladies in League Against Communism (LILAC), or the development-oriented Forest Protection Society bear witness to the diversity of goals, but uniformity of cohesion among community groups. This is not to suggest that these groups are in some way neglectful of the public good – merely that in tightly cohesive communities, (i) there may be an element of exclusionary behaviour, and (ii) not all community groups are the bastions of "warm fuzzies" as so often portrayed in the media.

The return of community has been interpreted by some academics as a return to simplicity, romanticism, traditional values, local knowledge and populist agendas. A few even go so far as to wish a return to the past: "forgetting the repressiveness, narrowness, parochialism, and real difficulty of life in earlier times, such simple nostalgia offers an attractive alternative to anxiety about the complicated present".[8] *Community* as used in this book does not naively refer to an emancipated, historical rural or village lifestyle freed from the harshness of the landscape. Nor does it refer to a static, geographical feature of the countryside. It is, however, time-dependent and will exist only for as long as the principles which led to its formation, maintenance and survival, are present. A sense of community will not spontaneously erupt or exist independently of the set of conditions which herald this nebulous concept. In the increasingly global culture which drives us steadily into the 21st century, community is not likely to be a simple, unified, cohesive subject – if it ever has been. Three off-shoots of this concept provide an important backdrop to the practice of environmental stewardship: sense of community, community development and community involvement.

Sense of community

Studies of community and communities have abounded in sociology and geography for many years. But, the concept of a *sense of community* was initiated by psychologists who saw the need to go beyond the study of individuals only, to study groups and communities undergoing structural and societal changes.[9] Community psychology has been recognised as a legitimate field of inquiry since the 1960s, but most studies of *sense of community* have taken place within an urban context in the United States and have concentrated on poor or disadvantaged residents. Much of the community psychology literature has focused on studies of social problems rather than environmental and economic concerns. The concept of *sense of community* is nevertheless relevant to the way that members of stewardship groups such as Water Watchers, Downside Landcare and the Mitchell River Watershed group relate to each other and their environment.

Solidarity based upon territorial ties with the environment is one of the basic notions of a *sense of community*. It is frequently argued that a sense of belonging, membership or identity within a community is also of paramount importance to this concept. The idea that any individual can influence a community group is also important, as is a shared emotional connection. Thus, while on the one hand it seems that *sense of community* is another catch-all academic phrase that has been used too often in the psychological literature, on the other hand, concepts of solidarity, membership and involvement, when examined individually, can explain how a *sense of community* is nurtured in community groups. In this way, it is a precondition for environmental stewardship. Consider one of the older definitions of *sense of community*:

> The perception of similarity to others, an acknowledged interdependence with others, a willingness to maintain this interdependence by giving to or doing for others what one expects from them, the feeling that one is part of a larger dependable and stable structure - these are some of the ingredients of the psychological sense of community.[10]

Sense of community is important because it is the principle which contributes most to concerted, participative stewardship action by local community groups. What makes up this principle - membership, influence, emotional connection and reinforcement of needs - form the basis for discussion about the three stewardship groups" *sense of community* in Chapter 6.

Community development

Community development has its roots in agricultural extension practice both in developed and developing world nations. However, it also has a strong history in the fields of community health, social welfare, adult education and community psychology.[11] As a profession community developers assist with difficult transition experiences for individuals arising from the inadequacy of existing institutions to cope with the problems of the times (land degradation for example). A set of common themes of community development practice can be identified as:

- empowerment (self-help);
- participation and involvement;
- social learning;
- mobilisation of resources;
- technical assistance;
- needs assessment;
- social justice.

Many of the goals of community development projects include some form of social change, whether they be focused on economic development; rural and agricultural development or health and education. There are three types of social change strategies within community development practice:

- providing people with sufficient information so that change is rationally justified;
- having people change their basic values and attitudes so that change is a process of re-education;
- forcing people to change using coercive power (usually within groups).

A critical issue here is the assumption behind these strategies that one group of people (government officers, welfare workers etc.) know what is best for another group (poor people, dislocated people, immigrants etc.). Fundamental clarification is required about "development for whom? development of what?" There is a crucial distinction between *development of* the community compared to *development in* the community. Like the current debates on globalisation, community development studies have raised questions about whether it is a process, a program, a method or a movement. Indeed, the whole notion of development, whether it is about sustainability in environmental terms, security in terms of local and global economies, or human health and wellbeing, has taken a battering:

> The more varied and contradictory the definitions of development become, the stronger its connotation. Development implies that one has started on a road that others know better, to be on one's way towards a goal that others have reached; to race up a one way street. Development means the sacrifice of environment, solidarities, traditional interpretations, and customs in the name of ever-changing expert advice.[12]

Rather than debate the relative advantages and disadvantages of community development any further here, some of its themes (empowerment, resourcing, provision of technical advice and social learning) are discussed in more detail in Chapter 6. Instead, given the rise in numbers of community groups undertaking stewardship activities and the swelling population of people involved in community-based environmental management, it is timely to examine the concept of community involvement.

Community involvement

The terms *public participation, citizen involvement* and *community consultation* are frequently mentioned in government reports and consultancy documents as examples of the increasingly common practice of community involvement, but each has a slightly different slant. In Arnstein's "ladder" of citizen participation, community involvement is measured along a continuum between persuasion at one extreme, and self-determination at the other.[13] Government initiated activities tend to emulate the former while self-determination is often the goal of a community group. In government circles there is more effort put into reducing low community involvement as an obstacle to grassroots action than championing the successes of voluntary participation in environmental stewardship. The result: local, *voluntary* involvement in this field has been neglected at the expense of "ways in which local people can be persuaded to provide the necessary labour input into environmental management projects designed outside the community".[14] This is desirable according to some Australian bureaucrats (see Chapter 5), but it is not the majority of opinion.

Other public participation factors more conducive to

improving government-community relations over local environmental stewardship have been determined by the Institute for Participatory Planning:

(a) The process of public participation should be agreed between the agency and participants.
(b) Public participation should start early in the decision-making process.
(c) The objectives of the public participation need to be clearly stated.
(d) People need to be aware of the level of power being offered.
(e) Efforts should be made by the agency to identify all interested parties.
(f) Information should be freely available to all participants.
(g) Participants should know how their submissions will be processed.
(h) Where appropriate (eg for travel and large time commitments) costs for participants should be reimbursed.[15]

From a stewardship group perspective, it is important to remember that commitment to the *process* of attracting involvement should not overshadow commitment to the task at hand. Indeed, many such lists as the one above have taken on levels of theoretical importance far outweighing any practical application. This raises the issue of whether public involvement is a means toward, or an end of, environmental management. The two are subtly different processes that may attract different types of participants.

More than just identifying who wants to be involved, it is critical to note the diversity and extent of people's participation. A broad "social reach" will mean that many different sectors of the community will become involved in direct and meaningful ways at different phases in the life-cycle of a project. And if the project leaders reach out to the community early on in a group's formation or a project's development, they will have a better chance of facilitating self-empowerment rather than disillusionment for its members. In other words, when public involvement is encouraged at the conceptual or design stage of a community project, the outcomes are quite different to that kind of "public involvement" which permits community groups to raise objections to a project at the end of the design phase. The superficiality of the kind of public involvement strategy which requires only labour and information from the community is obvious; it does not really *involve* the public.

The question of why and how people participate is often glossed over, but it is important to understand what lies behind community support or condemnation for a project. Some people are motivated to participate in stewardship groups because of the ease with which they can attain skills, knowledge or competence and with the ability to mobilise resources for local projects. Other benefits include, making better decisions, solidarity with the group, democratic credibility and a sense of responsibility to the community and to the environment. Conversely, but not surprisingly, the costs of participation such as lack of time and family commitments are associated with less involvement in community organisations. These issues are also discussed further in Chapter 6. Without a detailed understanding of environment however, the setting for community-based environmental stewardship would be lost.

Environment

Literally, an environment is that which surrounds. Another common definition of environment is that provided by ecologists. This is amongst the earlier definitions: environment is "the sum total of all the external conditions which may influence organisms".[16] Many environmental psychologists would add that this includes built humanised landscapes as well as "natural" landscapes and ecosystems. Importantly in this context environment also includes people and all the socio-cultural and historical legacies that we bestow on local places. Whatever the definition, for each of us as global citizens, what are perceived as being problematic or important features of local environments provide the context in which environmental stewardship will take place. For example, grasshoppers were perceived as being a distinct environmental problem for several government officers working with Water Watchers, because they caused premature tree-deaths in areas in which a lot of trees had been planted by State government departments. For members of the Downside Landcare group, dryland salinity, over population or capitalist agricultural systems were raised as serious problems for the environment. Obviously the environment and its problems are construed differently by different people and gain significance in the eye of the beholder.

The environment is not a blank slate upon which we humans impose our collective cultural wills. Nor is it something that we can only perceive (as useful, beautiful, dangerous etc) through the eyes of our culture and social organisation. Instead, the environment is both "naturally" and socially constructed. It is not a fixed thing waiting to be discovered, rather, an evolving thing that we come to know through increasing contact and interactions. Nature and culture, people and the environment are not at different ends of yet another set of dichotomies, but are mutually defining elements or agents of the same planet. The environment is larger and more comprehensive than a simple compilation of natural resources, human populations and architectural landscapes. Therefore, it is appropriate that the focus here is on *community-based* environmental stewardship - not natural resource management, after all it is the unique combination of special people acting in special places which give rise to local stewardship practices.

Humans act upon and in the environment, and the environment acts upon and shapes human perceptions. Environmental psychologists concerned with human-environment relationships and the effects of large cities on human psyches in the early 1970s now emphasise the stress and risk associated with potential environmental threats.[17] The field of cultural geography also recognises the effects of environmental characteristics upon the inhabitants of these environments - over time and as publicly recognised through metaphor and symbol. Environmental qualities such as dryness, heat, redness and expanse must shape the experience of the Mitchell River groups' stewardship of their catchment. The concept of environment as metaphor goes a long way toward informing the ways in which nations have historically constructed their environments.

Environment as metaphor

The countryside is an important concept in English rural localities, signifying England as a peaceful and tranquil series of hedgerows complete with moles and toads and other characters hopping in and out of children's literature. Indeed "the country" is a good source of signifiers of nationhood and nationalities all over the world. But nowhere has the "country" come to be more identified with a nation than in the bush of Australia. Akubra hats, out-back pubs, isolated beaches, the red centre, sheep, drovers and their dogs, have all helped shape Australia's rural identity and focus international attention on what it is to be from "down-under". However, despite the coherence of this rural vision for overseas visitors, the realities of life for inland Australians are diverse and refracted. Similarly the American West is not bound by any one set of generalised definitions whether in terms of frontiers, aridity, plains nor Federal land ownership. The only point of similarity between sub-regions of America's West such as Alaska, Northern New Mexico, Western Montana or the Rocky Mountains is a point on a compass.[18] To summarise, life in "the country" whether in the United States or anywhere else, like talk of "the community", it is not experienced in the same way by all members of rural society.

Despite a growing recognition that the Aussie outback is not merely comprised of the natural, wilderness, tropical or coastal post-card scenes most often associated with Australia, rural environmental issues still dominate public agendas. The environment of cityscapes, housing estates, urban parks, shopping malls, heritage buildings and other built forms has yet to capture the public imagination of city folk with sufficient strength to stimulate many suburban stewardship projects. This is notably not the case in North America where there is a longer history of urban green groups planting community gardens, undertaking heritage restoration work and conducting a variety of art and environmental projects to name a few examples. However, both rural and urban community groups provide an excellent opportunity to understand how people have historically constructed their environments as metaphors.

The environment as metaphor is a rich source of inspiration when looking at Water Watchers, Downside Landcare and the Mitchell River group. Common to the work of some landscape ecologists and geographers writing from the 1930s through to the 1970s was the belief that landscape, like the environment, is recognised as an important but fundamentally ambiguous term which may be "read" in a variety of different ways at different times and places and for different purposes by different people.[19] For example the same scene can be interpreted in many different ways: as nature, habitat, artefact, system, problem, wealth, ideology, history, place or art. Common to all interpretations is that these landscapes are socially constructed – leaving "clues" as to the type of people and cultures who inhabited them. In an Australian context, Frawley has classified five different, but sequential, cultural metaphors through which to view the bush – as a scientific, romantic, colonial, national and ecological resource.[20] These provide a particularly good framework for the way in which previous generations of Australians have viewed "the bush".

The scientific lens is an old one with reference to the Australian landscape. Joseph Banks was the first European to classify the environment from a botanical perspective and various physical and biological scientists have been trying to understand it ever since. In their quest to monitor phosphate levels in the creeks and streams in the Serpentine Jarrahdale Shire some Water Watchers group members have adopted this "scientific" metaphor for their environment. Group members suggested that since water quality was an indication of ecosystem health, by tracking and recording the highs and lows in pH level and phosphate readings, the whole landscape could be thus measured and interpreted as a system:

> For such persons the landscape that others may see is only a facade which their vision penetrates to reveal a transect of intricate pulsating networks, flows, interactions, an immense input-output matrix ... It is an ideology that implies a faith in man as essentially omniscient; that man through the rigorous disciplined power of his mind will eventually understand all that lies before him in the landscape; that ultimately through science we shall know the truth.[21]

This view of the environment drives many of the "monitoring" type stewardship groups such as those involved in bird and bat surveys, macroinvertebrate analysis and many of the -watch groups such as Frog Watch and Watertable Watch.

The *romantic* metaphor for the environment is also firmly ingrained in Australian history and self-identity. Painters from the Heidelberg school in the 19th century renounced paintings of seaside retreats reminiscent of more English scenes in favour of painting the Australian bush. It is ironic to note however, that the landscapes these painters depicted were used first by indigenous peoples then settled by Europeans and were therefore far from being "rustic".[22] In the second half of the 19th century, Australian "intellectuals" were encouraged to dispense with the romanticism of colonial impressions of Australia and emulate the American realist tradition of painting or writing about our own life and environment. Into the 1940s and 1950s the romance of the outback still set the scene for literary works of adventure, discovery and lyrical accounts of *Bush Christmases*[23] – how different, exciting and novel Australia was! Some of this romance and attachment to simple lifestyles is apparent in the current trend for traditional styles. Country-look clothing outlets, hewn wood furniture manufacturers and local craft galleries have capitalised on this romantic harbinger of safety and unity. Ironically though, the return of a new ruralist myth has ignored "the mass of misery, human decadence, and violence which once made the countrysides the places of "savagery" described by so many writers in the 1800s".[24] Emancipated or simply bourgeois, the trend in cottage fashion is to furnish our homes with blue and white checked cotton tablecloths and free-range eggs in baskets on wooden sideboards, while we sit in urban comfort and lament the fact that farmers are still clearing the last "virgin bush" off some rural properties. Are we still romanticising our environments?

The *colonial* lens was also popular for a long time in Australian history and, as much as I would like to announce its demise, in outback and isolated Australia, the twin practices of decrying the landscape and taming its harshness are still in vogue. Although pioneering explorers believed much of the Mitchell River watershed to be a rich and bountiful source of agricultural production, in the 1990s those of a development mindset like Harry claim that: "the country is useless – it wouldn't feed a mangy goat". Therefore, what better function for it than to rearrange it for human accommodation, to "improve" it, to erect artificial lakes and dams upon it and sell it off to unsuspecting but rich foreign landowners? This echoes another of Frawley's metaphors, that of *national* vision, confidence in Australian potential and pride of achievement which was particularly evidenced in the Sydney 2000 Olympic games. Removed from the city lights, in a township like Mareeba in far north Queensland with 40 per cent unemployment and many social and economic problems – not least of which is racism – it is common to celebrate local successes in the pastoral, mining, agricultural and tourism industries which will help forge the great Australian dream. But, not all members of the Mitchell River group share this highly developed nationalistic and colonising view.

The *ecological* metaphor is common in stewardship environmental groups. As public awareness and perceptions of environmentalism have grown, and with the dawning of widespread recognition that we are living off our "natural capital" at the expense of our environments, a public shift in thinking has occurred. Just as other social values (eg attitudes to family and fitness) have altered over the last 25 years, so too has public thinking about environments. But there has been a shift in direction. Choices around sexual preference, drug abuse and family composition have moved from being publicly validated decisions to the private realm, whereas environmental values and choices have moved in the other direction from the purely private (I'll do what I want to on my land) to the public sphere. The growing fields of environmental ethics and ecological economics are evidence of this change.

The ecological metaphor is perhaps the most important of the five described here in relation to stewardship groups. As potential ecological threats have emerged in local environments, there has been increasing public involvement in community groups all over Australia and internationally to help "combat" the damage. In turn this involvement in local environmental issues has fuelled the demand for restoration of ecosystems world-wide. Most importantly a focus on stewardship has facilitated the re-enchantment with independent, self-reliant and sustainable communities located in local places.

Environment as place

Sense of place creates both a oneness or connection with the environment and with the people and local organisations who inhabit those places. It is not something which is acquired when you move into a new area or suburb, it has to be cultivated. Like culture, sense of place is changing and adaptive, not primordial and fixed. And just as sense of place is not temporally confined, so it is not geographically limited to any specific rural or urban landscape. It has to be built in the minds of the beholders and the dwellers in a particular place, and it has to built over time:

> To arrive in a new and alien land, to build houses and towns, to join them with roads, even to clear the land and start to farm it, all this – the physical inhabiting – is much easier than the psychological settling of the country.[25]

Yi Fu Tuan has contributed much to the discussion of people-environment relations in terms of local identity with place. He coined the term "topophilia" to mean emotional attachment to and significance of local places.[26] Subsequently coverage of indigenous concepts such as belongingness, country and their interconnection have taken this literature on place attachment even further.[27] The extent to which place features in the relationship between community groups and their local environments is discussed in Chapter 6. In the context of environmental stewardship, there is another concept emerging - bioregionalism - which is gaining favour as a means by which local communities can live and work in local places while acknowledging the special significance of their surrounding environments.

Environment as bioregion

Some would argue that the history of bioregionalism is as old as civil society itself, originating from the Greek word *bios* meaning life and the Latin word *regio,* meaning territory under rule.[28] Bioregional theory has gained popularity recently and has become a recognised alternative among the many options presented under the banner of sustainable development. The essential idea behind bioregionalism is that boundaries of places should not be defined by administrative, political or other human constructs independent of the environmental context in which they are located. The concept has been defined in various ways corresponding to "natural" boundaries such as water catchments or social boundaries marked by the people who live and work in specific places.[29]

Bioregionalism owes much to the idea of a land ethic so eloquently expounded by Aldo Leopold (1949) in his famous *A sand county almanac*. It deserves repeating here.

> **The community concept**
>
> All ethics so far evolved rest upon a single premise: that the individual is a member of a community of interdependent parts …
>
> The land ethic simply enlarges the boundaries of the community to include soils, waters, plants, and animals or collectively, the land.
>
> This sounds simple: do we not already sing our love for and obligation to the land of the free and the home of the brave? Yes, but just what and whom do we love? Certainly not the soil, which we are sending helter-skelter downriver. Certainly not the waters, which we assume have no function except to turn turbines, float barges, and carry off sewage. Certainly not the plants, of which we exterminate whole communities without batting an eye. Certainly not the animals, of which we have already extirpated many of the largest and most beautiful species. A land ethic of course cannot prevent the alteration, management, and use of these "resources", but it does affirm their right to continued existence, and, at least in spots, their continued existence in a natural state.[30]

Whereas bioregional theory focuses on the land, and human interaction with it, Aldo Leopold, alongside indigenous peoples from many nations, considers humans to be merely another part of the whole community – natural as well as cultural, and therefore not giving precedence to either end of this artificial dichotomy.

Since the 1985 publication of Kirkpatrick Sale's *Dwellers in the Land: The Bioregional Vision*, various interpretations of bioregional philosophy and practice have emerged. Bioregional theory, without acknowledging the centrality of Leopold's ideas, cannot progress beyond the definition of the bioregion itself, unless it recognises and builds upon some "first principles". Kirkpatrick Sale identifies these as; community, conservation, stability, self-sufficiency, cooperation, interconnectedness, sense of place, cohesion, integration, decentralisation, complementarity, diversity, symbiosis, evolution and division. Despite being universally attractive as core human values, these principles cannot be universally interpreted as key elements of bioregional theory since they mean different things to different people. In practice, these principles serve to highlight a fundamental divide in the literature on bioregionalism between environmental determinism (ecological "fact" as precursor of bioregional value) and post-modern relativity ("facts" are socially constructed and have no primordial value outside socio-cultural communities).[31] The main difficulty characterised by this debate is the definition of a bioregion which assumes that some criteria, whether they are biophysical or sociocultural, be given primacy. Another problem mentioned in this debate on bioregionalism is a fixation with opposing forces be they biophysical or sociocultural phenomena. Such dichotomies are less than unhelpful. They actively promote dualism and other aspects of western instrumental rationality.

Bioregional theory potentially offers a great deal to the notion of place-centred environmental stewardship, but does not go far enough in working out how sustainable bioregional communities will act in the context of complex, chaotic, environmental degradation of the kind witnessed daily in local places. There is an immediate need for integrated approaches to (as well as a common language for) community-based environmental stewardship in the face of such complexity and uncertainty.

The context for integration

Widespread environmental uncertainty and complexity have given rise to new kinds of "eco-messes" and turbulence on an unprecedented world scale. Concurrently there has been a growing recognition of the need for new problem solving approaches which integrate nature-culture studies with work done "on-the-ground" in both the socio-cultural and biophysical realms. Coupled with increasing complexity in global environments is the notion of interdependency in the social, financial and cultural spheres. "Unravelling" tangled webs of inter-relationships and causal linkages is neither possible nor desirable. This complexity is almost always apparent in many of the contexts in which stewardship groups act. Given that there are very many problems, there must be many attempts at solutions, on many different scales and across theoretical and practical domains. Proponents of community stewardship argue the case for developing local solutions to local problems. But as Tim, a university lecturer associated with the Downside Landcare group reminded me, "we must get away from the idea that everything is knowable. Not all problems are solvable - often the best we can hope for is to improve situations".

To operate in this context of complexity and uncertainty has called for the development of "post-normal science", that is, an extension of traditional science, appropriate to the conditions of the present age. Scientists who propound post-normal science argue that principles of uncertainty and ignorance can no longer be expected to be conquered, instead, they must be managed for the common good. Post-normal science responds to high levels of environmental uncertainty and takes place within a context in which the decision-making and policy stakes are also high. Instead of solving problems, perhaps it is now more appropriate to think in terms of "coping with and ameliorating" complex issues.[32] Although post-normal science has been associated with to global issues and large scale environmental problems, I believe that it now has a place in environmental stewardship at smaller scales since there is no less uncertainty at this scale, it is just qualitatively different.

As the issues and problems of environmental management become more and more specialised, the answers become less and less relevant to the rest of society. Solutions will not be found merely by labelling the complexity of the Mitchell River Watershed as "an agricultural problem" or "a transport problem". This is actively misleading. What is necessary is "a critical recognition of the inevitable limitations of our own perceptions, and to integrate scientific knowledge with the many sources of social, economic, cultural and intuitive knowledge relevant to complex problems".[33] In addition to collaborative approaches, the environment also needs approaches which draw upon an eclectic array of practices, theories and methods as tools of explanation, not as ends in themselves. I am not suggesting that traditional academic single disciplinary approaches be discontinued, rather that environmental research should not be limited to specialist description and analysis. Everyday practices of environmental stewardship groups are equally valid and for sustainability we need multiple voices.

"Hard" scientists have long eschewed "soft" science approaches to environmental problems and vice versa. The arts and humanities have long held themselves aloft from the engineers, foresters and economists. It is almost as if every discipline speaks a different language and over time this has led to entrenched ideological suspicions between academic departments not to mention different government departments. Without delving into history, it would be true to say that there are many reasons for different departments and organisational groups to mistrust, ignore and shun each other, but it doesn't take the multi-faceted problems of environmental degradation any closer to amelioration, if not solution.

The investigation of "situation-based" problems is an important reason to choose an eclectic, holistic and integrated approach to the environment. For all kinds of environmental activists, stewards and managers, it is important to describe and analyse (i) context specific problems; (ii) from multiple perspectives; (iii) with an approach which can best be labelled "intelligent pragmatism".[34]

The principles of integration

In all this discussion of the need for integration and the importance of finding a common language, there are two critical themes. First, top-down and bottom-up approaches are both necessary, as are those of the natural and social sciences. But these simplistic dichotomies are not alternatives or mutually exclusive. They are merely heuristic devices used to provide a means of approaching this topic. Hence, rather than encourage a middle-ground compromise (although this may sometimes be necessary), I am suggesting that new ways of promoting environmental stewardship be constructed by establishing a means of dialogue and exchange between the respective ends of these continuums.

Second, this integration is not an attempt at universality; not everything has to be considered, weighed up and rationalised. Holism is not something one does. It is something to be considered when placing oneself along the many points of the practitioner – academic, top-down – bottom-up or left-wing – right-wing continuums to name just a few. Holism cannot be applied or understood as the science of adding up a series of fragmented, reductionist, discontinuous parts.

Figure 2.1 Model of integration

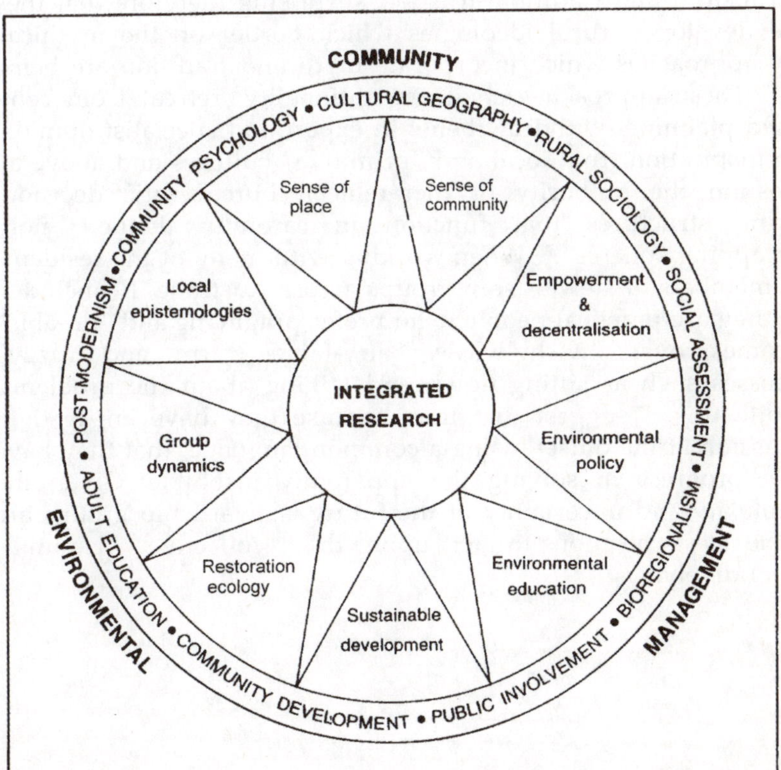

Figure 2.1 provides one model of stewardship. Around the outside are the various fields, ideas or disciplines which have contributed to the development of this common language. Note that the diagram neatly captures these concepts with hard definite lines separating what is in and what is outside the figure; but there are many other key ideas that could have been included. Not all of these are represented here, just the main ones. Also note that for the sake of visual appeal, all segments and categories of knowledge are represented as contributing equally to this topic. This is a false claim, but a convenient one for the sake of expressing the essence of these ideas diagrammatically.

Society can no longer unquestioningly rely solely on structures which are premised on science and rationality, especially when attempting to redress environmental, economic and social pathologies in rural communities. These places are firmly rooted in local history and tradition, it is not surprising therefore that they have developed rural ideologies which border on the mythical. New approaches which incorporate myth and tradition are being tried. These approaches challenge rationality, retreat from centralised planning, withdraw belief in expert and specialist opinion, gain inspiration from local and "primitive" cultures and above all "question the authority of hierarchical, bureaucratic decision-making structures that function in carefully defined, non-overlapping spheres".[35] When working with many of the residents and members of stewardship groups across Australia, I concluded that they are practical people, who prefer pragmatic and "do-able" outcomes over "wishy-washy, airy-fairy warm and fuzzy" processes such as sitting down and talking about the problems. Despite this, I suggest that it is because they have engaged in discussion at the outset, using a common language, that they have made progress in solving the apparently insoluble. Given the complexity and uncertainty of the future, stewardship groups are in the best position to act upon their collective wisdom – cooperatively.

Endnotes

1 For Australia the insertion of "ecologically" before "sustainable development" seems to have made little, if any, difference to its implementation as a practice not merely as a desirable policy goal. For a critical overview of the concept, see Hamilton and Throsby's 1998 edited collection.

2 Lerner 1993: 4; 1994. Also see Ordubegian 1993, Gardner 1991a and Knight and Landres 1998 for an overview of research into environmental stewardship. The stewardship theme is premised on original work by Leopold 1949, republished in 1966 and is expanded in Roberts 1990.

3 See for example, Robin's 1998 book *Defending the Little Desert* where she discusses links between soil conservation, environmentalism and stewardship (p150). For a broader discussion of this topic, see Roberts' 1990 collection of papers entitled *The birth of LAND CARE*, especially his essay on Land Ethics where he speaks of a theology of the land and Christian land ethics.

4 Steelman and Carmin 1998: 496, 497 talk about stewardship in a study conducted in the Laurel Mountain community of West Virginia.

5 Hillery 1955.

6 Condon 1991: 3.

7 Sarason 1974: 157.

8 Leinberger and Tucker 1992: 87; see also Davidson 1993.

9 See for example, Sarason 1974; Levine and Perkins 1987; Heller 1989, 1990; Grierson 1993; McMillan and Chavis 1986; Newborough 1996 and Dalton, Elias and Wandersman 2001. These authors offer a very good overview of this topic from its inception to the present. In particular, the *Journal of Community Psychology* is a good place to track its development.

10 Sarason 1974: 157.

11 For a good introduction to the community development literature, see; Summers 1986; Korten 1980,1990; Phifer, List and Faulkner 1989; Heller 1989; Blakely 1989; Christenson 1989; Christenson and Robinson 1989; Wilkinson 1989; Kretzmann and McKnight 1993; Lappe and DuBois 1994; Ward 1994; Ife 1995; Checkoway 1995; Kaye and Wolff 1997 and Dalton, Elias and Wandersman 2001.

12 Esteva 1987: 144.

13 Arnstein 1969.

14 Ghai and Vivian 1992: 3.

15 These factors are reproduced in whole as they appear in Syme 1992: 71. Other useful and more recent resources include Waters' 2000 *The community planning handbook: how people can shape their cities, towns and villages in any part of the world,* or for an Australian analysis of the

participation literature see Buchy and Ross 2000 *Enhancing the information base on participatory approaches in Australian natural resource management.*

16 Hanson 1962: 126.

17 Ittelson, Proshansky, Rivlin and Winkel 1974 commented on this subject a generation ago. Now see Evans and Cohen 1987, Graumann and Kruse 1990, or Hobfoll, Briggs and Wells 1995 for more up to date work related to this topic.

18 Flores 1999.

19 See Lewis, Tuan and Jackson in Meinig's 1979 edited collection for a rich introduction to this field.

20 Frawley 1992.

21 Meinig 1979: 38.

22 Gammage 1994 and Johnson 1994 comment critically about this misconception.

23 Gathcart-Borer and Smart 1947. See also the 1947 and 1983 films of this Australian classic to appreciate how the Directors interpreted the romance of the Australian bush.

24 Karnoouh 1986: 11.

25 Carroll 1982: vii.

26 Tuan 1974.

27 See for example Deborah Rose's *Nourishing terrains: Australian Aboriginal views of landscape and wilderness* 1996 or Peter Read's *Belonging: Australians, place and Aboriginal ownership* 2000.

28 Gardner and Roseland 1989.

29 It has had a longer history in the United States and Canada than in Australia and New Zealand as evidenced by Sale's 1984 and 1985 publications. See also *Bioregionalism* edited by Michael Vincent McGinnis, 1999. This book offers an excellent update on bioregional approaches to the study of home places, regions and globalisation, local knowledge, modern science and the future!

30 Leopold 1966: 219.

31 See Alexander 1990 for a more comprehensive discussion of this split.

32 Funtowicz and Ravetz 1991: 146. Although there have been many subsequent reincarnations of Funtowicz and Ravetz famous typology, to understand the topic as it was introduced at the time warrants reading the original.

33 Carley and Christie 1992: 155.

34 Walker 1987.

35 Rosenau 1992: 7.

3

Stewardship groups: in profile

Scale and boundaries

Stewardship groups form for different purposes at different societal and geographic scales. One of the most well known rural community groups in Australia is the Country Women's Association which is present at both a national and local scale and has a well-defined political constituency. Another famous Australian rural organisation is that of local bush fire brigades, which operate locally, but have a State-level statute and work across State boundaries in times of national crises. Stewardship issues are similarly covered at different scales. In the United States, volunteer environmental restoration, stewardship and advocacy groups have formed in thousands. For example, Friends of Five Creeks undertake stream daylighting at a local level, whereas the Greenbelt Alliance is a regional scale advocacy group which protects the Bay Area's farmlands and open space. Both operate in and around San Francisco. There are very many more such organisations. Knowing and identifying the scale at which stewardship groups work helps to understand the parameters of their work.

The geographic boundaries and hence scale of activity for stewardship groups are determined by a mixture of social, political and biophysical considerations. Their focus is usually local, but they may also contribute to regional initiatives. Sometimes they work at a national level and occasionally at an international scale. Water Watchers used shire boundaries to define their area of interest and operated locally at both geographic and societal scales.[1] The Downside Landcare group is based on social as well as catchment boundaries. Landholders from neighbouring properties who meet socially at church or the tennis club for example, are more likely to belong to a landcare group compared with those on the outskirts of the sub-catchment who do not meet up on such a regular basis (see Chapter 6 for further discussion on this). Although geographically the Downside Landcare group operates at a local catchment and sub-catchment scale, it is linked "by deed" with hundreds of other landcare groups recognised throughout Australia. The Mitchell River group on the other hand operates at a vastly bigger scale and is defined solely by its catchment boundaries, although these are far from distinct in many places. Like much of Australia, the catchment is largely flat, but unlike many parts of the country the presence of the mighty Mitchell River (which starts as a trickle and ends in a majestic tidal river delta where it meets the Gulf of Carpentaria) serves to unify the group. Incongruities between natural and human system boundaries are a deep-seated problem and present many challenges to those with responsibility for implementing ecological sustainability within a local area.[2] Therefore it is important to be clear about the scale at which various groups operate, see Figure 3.1. The three stewardship groups I visited operate not only at different scales but in very different parts of Australia, see Figure 3.2. Each of the three groups attract a slightly different membership type. But perhaps more interesting than a demographic overview of education and income statistics is a profile of a few typical members of each group. Throughout the book I have used pseudonyms to refer to particular individuals.[3]

A few definitions

Stewardship groups can be loosely classified as the environmental equivalent of other community groups. To understand the former it is educative to briefly examine the latter. The concept of *community group* suffers the same lack of specificity in definition as does the more generic term *community*, discussed in Chapter 2. With the advent of sustainable development as a policy tool, community groups are being seen by parts of government bureaucracies as convenient administrative units destined to implement that nebulous goal. Thus *community group* adopts an ambiguous definition and may simultaneously be used to convey an administrative unit and a social reality constructed of insiders and outsiders. Canadian research further distinguishes between advocacy groups and stewardship groups.[4] These two terms describe the ends of a continuum rather than separate categories. Where advocacy groups are more active in lobbying for the environment, stewardship groups have a greater action-protection role to care for some aspect of their local environment. Advocacy groups undertake surveillance or watch-dog work to record and

Figure 3.1 Stewardship scale

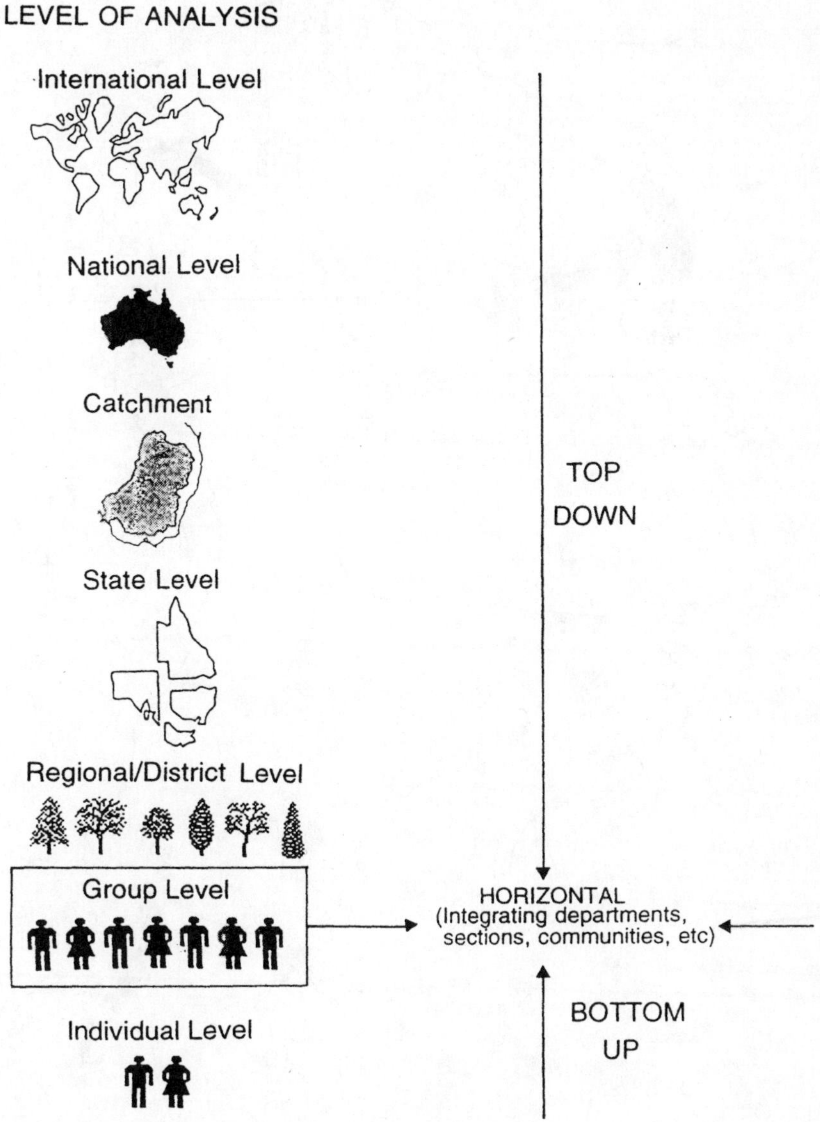

Source: Adapted from Chamala and Mortiss 1990

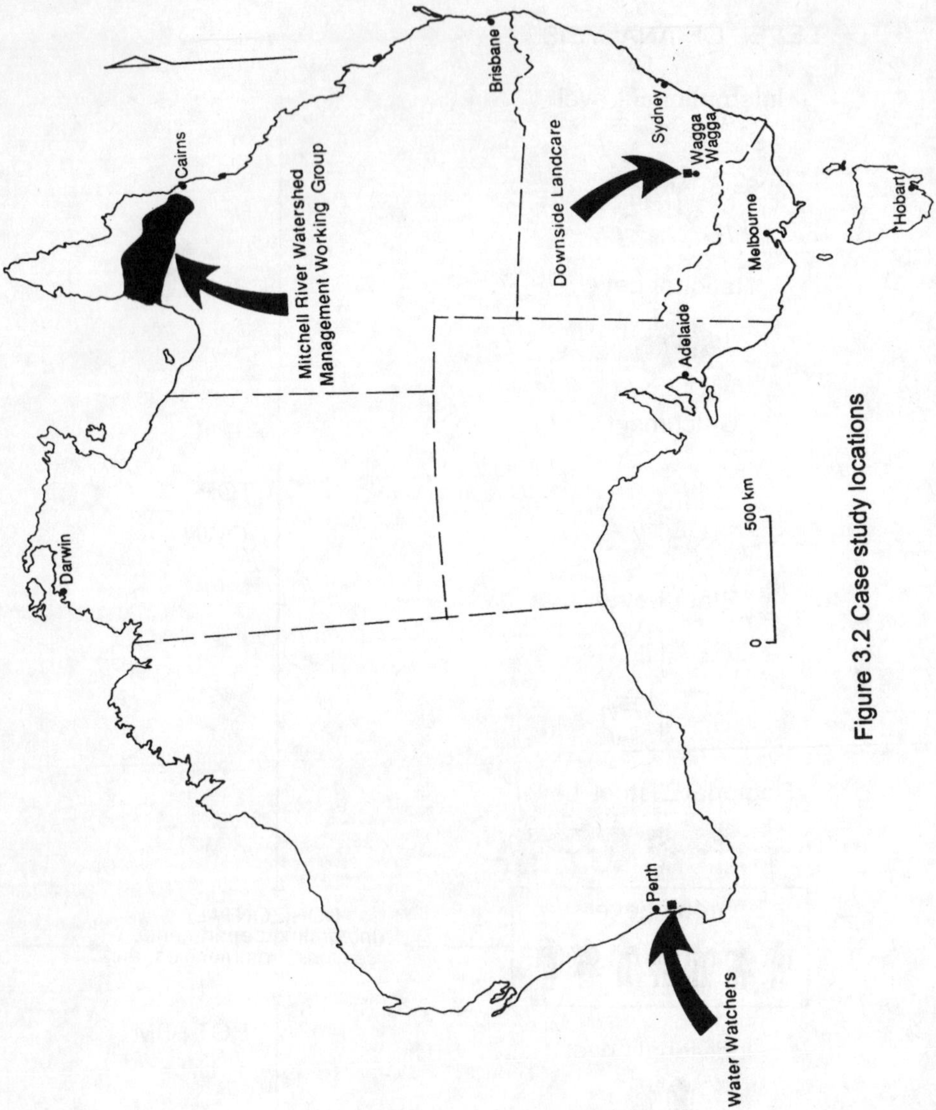

Figure 3.2 Case study locations

report environmental vandalism, pollution, poaching and breaches of environmental protocol in development projects. They tend not to form locally and often operate at national and/or international scale. They are not the central concern of this book. It is the local stewardship group, alias water quality monitoring group, landcare group and catchment management group (among others), which is the current focus of much international attention to restoration ecology and participative processes. While stewardship and advocacy are generally well comprehended, restoration is still fairly new to the public. It is defined as follows:

> Ecological restoration is the process of assisting the recovery and management of ecological integrity. Ecological integrity includes a critical range of variability in biodiversity, ecological processes and structures, regional and historical context, and sustainable cultural practices.[5]

Stewardship groups derive their constituency from segments of wider society, for example organisations of farmers, or graziers. What is more interesting is how stewardship group members like Robyn of Water Watchers, Ellen of Downside or Macka of the Mitchell River group define themselves in terms of the aims and objectives of their group (Box 3.1).

Box 3.1 Stewardship groups defined

- A group of people working together for a common goal through common achievements to rectify problems.
- People within a defined area working toward one cause.
- A group that involves the community – run by the community rather than driven by someone else.
- People who work together to get something done that needs to be done.
- Socially stimulating, I mean convivial, a lot of conversation not all serious business, fun too. We can gain ideas by getting together and exchange ideas and problems and solutions to the problems.

Source: Interviews with group members

However, mere definitions of stewardship groups do not explain what they do. Box 3.2 summarises the main restoration activities stewardship groups undertake and gives Australian and Canadian examples. Arguably these two countries have the most advanced stewardship programs and longest history of voluntary environmental stewardship in the world due to their system of governance, low population base and geographic expanse.

Box 3.3 illustrates other types of stewardship activity that community-based environmental groups often engage in, and also draws on Australian and Canadian examples. Box 3.4 establishes the common attributes of a voluntary or community association.

Box 3.2 Types of restoration activities

ACTIVITIES	**AUSTRALIAN and CANADIAN EXAMPLES**
Enhancement	
Any type of activity designed to improve an environment or species habitat to facilitate environmental restoration. Examples include building and maintenance of nesting boxes, spawning areas; or development of vegetation projects and trails.	Narrandera Remnant Woodlands Enriching Biodiversity and Ecosystem Persistence (AUS) Wildlife Sanctuary for Captive Breeding of Endangered Species (AUS) Emu Swamp Koori Landcare Bush Regeneration Wildlife Corridor (AUS) Regeneration of Native Perennial Grasses Hard Red Mulga Country (AUS) The Bruce Trail Association (CAN) Glen Morrison Dung Beetles for Enhanced Water Quality (AUS) Increasing Native Grasses Through a Soil Management Strategy (AUS) Upper Doughboy Creek Fencing and Revegetation Project (AUS)
Rehabilitation	
Any type of activity designed to repair a degraded environment or rehabilitate a particular species. Typical activities include clean-up projects, riparian zone repair, tree planting, reintroduction of species, erosion control, weed control, feral animal control.	West Pennant Hills Blue Gum High Forest Rescue (AUS) Clean up of Driftwood by Volunteers on Boundary Bay, British Columbia (CAN) Medicine River Wildlife Rehabilitation Centre (CAN) Urban Habitat Restoration: Long Nose Bandicoot Population (AUS) Friends of Lily Lake (CAN) Revegetation at Historic Bow Wow Creek Gorge (AUS) Restoration of Degraded Banana Lands Cudgen Nature Reserve (AUS) Crescent Head Little Nobby – Surf Club Erosion Control (AUS) Goonoo Goonoo Stream Bank Restoration Project (AUS) Emu Swamp Drainage Watercourse Repair and Revegetation (AUS) Matheson Chilean Needle Grass Control Program (AUS)

Preservation / conservation	
Any type of activity designed to preserve or maintain certain environmental values within an area. Activities emphasise the importance of not allowing key habitats to decline (any further) and/or protection of endangered species or wilderness areas through designation of an area as a park or protected place.	Bruderheim Natural Areas Society (CAN) Bicentennial Park Wildflower and Orchid Protection (AUS) Bras d'Or Stewardship Society (CAN) Protection of Aboriginal Sites at Mystery Bay (AUS) Okanagan Region Wildlife Heritage Fund Society (CAN) Barlings Beach Archaeology and Habitat Fencing (AUS) The Islands Trust Fund – BC (CAN) Tanilba-Mallabula Walkway: Protecting Foreshore Dune/Koala Habitat (AUS) The Friends of Pukaskwa (CAN) The Hamilton Naturalists' Club (CAN)

Adapted from COMMUNITY STEWARDSHIP: a guide to establishing your own group *(Fraser River Management Program et al). Australian examples from Environment Australia* http://www.nht.gov.au – *(Natural Heritage Trust).*

Box 3.3 Other types of stewardship activities

ACTIVITIES	AUSTRALIAN and CANADIAN EXAMPLES
Planning and management	
Any activity which may address the long-term biophysical, cultural and social well-being of a region (often watershed based). Groups' planning and management efforts are not always integrated with the broader environmental context ie. restricted to a river, wetland or species.	Salmon River Watershed Management Partnership (CAN) Planning to Enhance Turquoise Parrot Habitat (AUS) Eurobodalla Coastal Environmental Capacity Planning/Coastal Open Space Planning (AUS) Western Riverina Natural Grasslands Plan (AUS) Ensuring Sustainability: Pomona Community Action Plan (AUS) Nicola River Watershed Round Table (CAN) Bow River Basin Water Council (CAN) Dumaresq Creek Catchment Management Plan (AUS) River Action Plans – Williams River Riparian Corridor (AUS)
Education and interpretation	
Any activity which addresses environmental education both in schools and the wider community. Activities include the establishment of nature education centres and interpretive trails, walks or exhibits.	Wirraminna Burrumbuttock Arboretum Environmental Education Project (AUS) YoWoChAs Outdoor Education Centre (CAN) Illabo School Bush Tucker Garden (AUS) Gulpa Creek Educational Walk Trail (AUS) Kingfisher Environmental Interpretive Centre (CAN)

	Indigenous Weaving Garden (AUS) Bennett Environmental Education Centre (CAN) Ash Island Mangrove/Saltmarsh Interpretive Boardwalk (AUS) Royal Easter Show Rangeland Education Theatre (AUS) Medicine Hat Interpretive Program (CAN)
Survey, assessment and monitoring	
Any activity which uses environmental indicators on a continual basis to monitor and assess specific species, states or conditions of local environments eg frog-watch, water-watch and fish counts. It may take place at local, regional or international scale in single community groups or over an entire international network.	Monitoring of Bacteria and Nutrients in Meads Bay (AUS) Nova Scotia Plant Watch (CAN) Grey Nurse Shark Migration Project (AUS) Eagle Count in Squamish Estuary (CAN) Aboriginal Water Testing and Quality Project (AUS) Beaverhill Bird Observatory Society (CAN) Checking the Mantle of Green (AUS) Parkland Airshed Management Zone (CAN) Echidna Watch (AUS) Friends of the Eltham Copper Butterfly (AUS)
Mapping and research	
Activities designed to conduct one-shot research inventories of species, land, air or water quality. Projects may include maps of environmentally sensitive areas or GIS profiles of land use and/or condition.	Nambucca Valley Estuarine Mapping and Resource Inventory (AUS) Identification of Ecologically Sensitive Areas in Surrey (CAN) Mapping Vegetation Types for Catchment Planning in NSW (AUS) Costs of Land Degradation in the Boorowa River Catchment and Downstream (AUS) Berowra Creek Water Quality Database (AUS)
Awareness and demonstration	
Any activity designed to raise community awareness. Projects include: a)publication of resource materials, videos, brochures or pamphlets b) distribution of information to the public via electronic, or broadcast media and c) demonstration sites, field days, open days, workshops etc.	Farm Forestry Demonstration Sites (AUS) Streamkeepers Handbook: A Practical Guide for Stream and Wetland Care (CAN) Rural Land Modification Awareness Project (AUS) Urban Total Catchment Management Community Awareness (AUS) Ecologically Sound Practices in Everyday Life (AUS) People and Planet Friendly Events (CAN) Botany Wetlands Public Awareness and Education Project (AUS)

Adapted from COMMUNITY STEWARDSHIP: a guide to establishing your own group *(Fraser River Management Program et al). Australian examples from Environment Australia* http: //www.nht.gov.au – *(Natural Heritage Trust).*

Box 3.4 Attributes of a voluntary organisation[6]

- Voluntary participation and flexible membership.
- Little internal specialisation, open to restructuring over time.
- Membership control over collective affairs.
- Loosely organised, non-hierarchical.
- Task oriented with flexible task allocation.
- Absence of clearly defined rank and status.
- Emotional attachment to task.
- Limited rules governing behaviour.
- Participation is generally limited to a small proportion of an individual's time.
- Membership recruited through common interests.
- Action learning/research approach.
- Frequently spontaneous and able to respond quickly.

Group formation

The way in which community groups form is subtly different from the factors which motivate particular individuals like Robyn or Mike or Lena to join these groups. Although the threat of some unwanted event or impact may provide the spark for individuals to seek group membership, it does not explain the original impetus for group formation. Groups do not appear because a collection of individuals simultaneously worry about sudden or incremental changes to their lifestyle, landscape or health. Someone has to initially take action upon the basis of their ideas, experiences, attitudes and beliefs and define the urgency of the situation. Mike of Downside, Jenny of Water Watchers and Vic of the Mitchell River group were those such individuals for each of these stewardship groups. But what drove them to do it? There is not enough known about how and why individuals decide to act to defend an aspect of their environment. Canadian research however, has established that there are some catalytic factors which aid in the process of stewardship group formation:

- contact with existing groups and their responses to environmental problems in other geographic contexts (both regional, State-wide and globally), through direct contact and visits, media coverage and electronic linkages;
- public fora which allow individuals or groups to meet and form coalitions, these may be chaired by representatives of governments, community groups, universities, etc;

- individual encouragement from politicians, academics, activists and public servants, this may include seed-funding and technical or other advice;
- splinter groups forming in defiance of, or in rebellion to, an existing group's attitudes or beliefs.[7]

Within an Australian context, there is an additional factor that influences group formation:

- Government policy that encourages or facilitates community group action on local environments.

The federally funded Natural Heritage Trust (NHT) of Australia had a huge impact on stewardship group formation. With the privatisation of the national communications company, Telstra, came proceeds from that sale which were politically directed to environmental groups all over the country. Some said it was a windfall. But like all windfalls, the produce, unless picked up, eventually turns bad. In 1992, 51 per cent of the 180 responses from a survey of landcare groups stated that a general concern about land degradation or a specific reaction to salinisation and water table problems were the factors which prompted group formation in their area.[8] A further 16 per cent said that landcare groups started on the basis of desire for action at the *grassroots,* a belief that landcare could help, a need for a more participative model or a concern about others capturing the green agenda. In 2000, environmental degradation issues were still captivating the hearts and minds of local people whether in landcare groups or other stewardship groups. But the crisis mentality and administrative burden associated with Landcare at the end of the Decade of Landcare has meant that although the issues still remain, the enthusiasm and initial momentum for landcare has waned. With NHT funds running out, groups are still forming, but at nowhere near the same rate as at the beginning of the last decade. These issues are taken further in Chapter 5, "Red Tape or Green".

Downside Landcare group

The Downside Landcare (DL) group was established and grew as a result of a bus tour to Victoria initiated by the New South Wales Department of Conservation and Land Management (CaLM).[9] Members of the Downside Agricultural Bureau witnessed severe dryland salinity problems and spoke with people involved with landcare groups in the Benalla region. Upon their return, a public meeting was organised by Mike and Ellen, one of the farming couples who had seen the neighbouring salinisation. Both husband and wife telephoned everyone who lived in the Downside area and invited them to attend. At that meeting, there was widespread agreement that a landcare group was necessary to address salinity problems already visible in Downside and a committee was formed.

The fledgling landcare group was supported by a lot of "cross-fertilisation" of members from the local Tennis Club (where meetings are still held and a nursery has been established) and a local Tree-planters group. These horizontal networks were, in turn, supported by vertical links to government policy through local Agricultural Extension Officers and the cooperation of two key government departments.

Membership of the Downside Landcare group is typified by broad-acre farmers enjoying middle-age. Daisy is a farmer in her late fifties, is married to a farmer, and has two adult sons. She holds many voluntary positions within the district and has a wide variety of interests. She is an elder in the local church where she is well respected and hard working. She does not play sport, has a secondary school education and is well-established financially. Mike is optimistic about landcare. Both he and his cousin Marjorie are active in the local tennis club and believe that if everyone were to band together in the community, salinity could be beaten. To that end, Hal and Marjorie are also members of the local Society for Growing Australian Plants and are concerned about the number of non-endemic tree species still being planted in the district.

Although Downside Landcare group's primary goal is to combat dryland salinity, in doing so members have also embarked on a more general level of community development. The Downside group have developed a stewardship ethic for landcaring prefaced on a sense of responsibility for present and future generations. They are proud residents of their district, focusing not simply on solving "the problem", but also on sharing their achievements and caring for the community. People like Mike and Ellen, Bobbie, Chad, and Kim (a group comprised equally of men and women) have developed landcare-related skills through involvement with the group, and have developed a strong commitment to working alongside government departments. Individual farmers have also made significant productivity gains through working together and sharing local agricultural knowledge.

The Downside Landcare group also started out with a strong local voluntary leader who supported the group through its establishment and consolidation phases. There have subsequently been four other chairmen. The founding chairman has gone on to represent the NSW Farmers' Federation on various catchment committees and salinity summits often by invitation from the Premier. When I last spoke to him, he told me that he was just about to ease himself out of landcare, environment and catchment management commitments ... but he has been telling me that for eight years! It also had a well motivated and energetic local facilitator who left the group after her two year contract finished. However, she did not leave the region and is now working with regional community groups on water reform issues. The Downside Landcare group then hired two more local people on short-term contracts to facilitate the group (one after the other), but quickly reached an optimal stage of development, which allowed the group to become less reliant upon government support. They have been independent of government since 1995, both financially and administratively. The group has supported its stewardship activities through recycling chemical drums (used on farms for pesticides, herbicides, fungicides etc) and by growing tree seedlings. Many of the landcare group farmers buy trees from the group at slightly higher prices than they could find elsewhere so they can feed money back into the group to upgrade the nursery. The Downside Landcare group has significantly transformed their two valleys in projects that have taken concerted action against dryland salinity and associated environmental degradation problems. On one property alone saline ground water levels have dropped from just above ground level between 1990 and 1995 to seven feet below ground level in 2000, and thousands of trees have been planted where before there were few. And it has not only been a landscape transformation. Personal and development opportunities have been created for and by the people who live in these peaceful valleys. Through all the changes to the office bearers of the Downside Landcare group there has been one unassuming, hard-working and committee landcarer who stands out from the others. In 1995, Billie the Publicity Officer's efforts were recognised when she was awarded the NSW Landcarer of the Year Award.

Water Watchers

Water Watchers (WW) began working as a women's group, but grew to encompass both women and men who live and work inland of the industrial area in Kwinana, south of Perth.[10] Members were primarily involved with monitoring phosphate levels in the streams, creeks and drains within the Serpentine Jarrahdale Shire. Jenny, Janean and Ken believed that local residents were being unfairly blamed for contributing toxic levels of phosphates to the waterways which in turn were causing outbreaks of blue-green algae blooms in a suburban estuarine inlet. They spoke of the need to counteract the blame they felt governments laid on the Serpentine-Jarrahdale community. Water Watchers formed largely in response to the threat of being labelled perpetrators of environmental destruction.

The group started informally around Jenny's kitchen table in Serpentine Jarrahdale. Jenny was a shire councillor at the time and invited an officer from the Office of Catchment Management (Colleen) to meet with herself and Louise who was new to the area and wanted to get more involved in the community. Each woman was interested in water quality issues (from quite different perspectives) and each was committed to raising environmental standards. Through Colleen's knowledge of funding sources, a grant application was submitted to the Office of Women's Interests. The group gained seed-funding as a women's initiative under the guidance of Louise who put up posters in the local shopping centre and acted as coordinator and contact person for residents interested in the project.

Membership of the Water Watchers group was typified by married women with children. Robyn was in her late thirties and regularly volunteered her time at the local school as well as teaching guitar after school hours. She ran a business with her husband and was concerned about being adversely affected by poor water quality. She and Louise were both concerned about the future lives of their children, wanting them to grow up in a clean and healthy environment. Janean was a professional radiographer, did not attend church or play sport and considered herself "middle-class". All these women worried that not enough people cared about water quality. The group was rounded out by Buster, an older man near retiring age who owned an organic vegetable growing business and took a very hard line on environmental stewardship – not least because he was worried about the effects of agricultural chemicals on his family's health.

Among their many achievements, Water Watchers may have been the first community group to produce water quality data which has subsequently been added to government databases. In Western Australia their efforts toward integrated catchment management have been recognised by two Ministers of Parliament and, more importantly, their work has established a precedent for community and government working together. The Water Watchers group received a great deal of support and advice from the Community Catchment Centre established and funded by government, and based nearby in Pinjarra; this is also detailed further in Chapter 5.

Water Watchers started out strongly with a well defined and locally active constituency. The group had not been in existence for very long before two key group members left to pursue other commitments. Nevertheless, the group fulfilled its major goal of monitoring local water quality over three consecutive wet seasons. However, the extent to which the remainder of the group was prepared to continue monitoring water quality fell dramatically. Members were no longer content just to monitor water quality ad infinitum. They had completed what they set out to do and located point source emissions of phosphorus into the waterways. Faced with a reduced membership and completed short-term goals, the group was confronted with three apparent choices:

(a) to disband,
(b) to continue monitoring in a different format – perhaps through the development of biological indicators to develop an inventory system with which to compare old data, or
(c) to regroup as an advocacy group, tackle the local political system and intervene to prevent point source emissions from contributing excessive quantities of nutrients to the creeks and drains in Serpentine Jarrahdale.

Although the group had strong vertical links into government agencies and finely tuned networks which catapulted it into the limelight in its establishment phase, it did not have the necessary horizontal ties in order to build the cohesion, identity and empowerment necessary for a sustained sense of community. In hindsight this rapid expansion may not have allowed sufficient time for extended community building and the development of rapport among group members. The group decided to disband, but their stewardship efforts have been far from wasted.

Skills and experience gained by some group members have lead to grant proposals being successfully funded in other environmental arenas. Hundreds of thousands of dollars are being committed to the Shire from the Natural Heritage Trust. Complementing the water quality monitoring projects, there is now a focus on improving the Serpentine River, restoring bushland, managing weeds, revegetating road-sides, seed-collecting, growing and planting local species. Out of the work of Water Watchers came recognition of environmental deterioration and the need to revegetate their Shire. They also made the scientific "discovery" that the phosphorus that was washing down the creeks and drains in Serpentine-Jarrahdale was not soluble, leaching out of over-application of fertilisers, it was particulate matter coming out of the heavy soils. It was therefore a matter of preventing soil erosion and that gave rise to renewed interest in stewardship in the Shire.

In 2000 there were 10-15 local stewardship groups working in the Shire – some of whom were spawned directly from the earlier effects of Water Watchers. There is a new meeting place for local groups to use in an ex-school situated near the Shire offices. And there are approximately 70,000 trees per day being planted in the Shire by a band of local volunteer stewards leading to a very pronounced landscape transformation. According to Jenny, Shire President, Water Watcher and well respected environmental steward:

> If anyone asks me why our Shire has its reputation for environmental concern and management, I would suggest it was begun by women who cared, and developed because of a belief that knowledge is the foundation for change and carrots work where sticks don't.[11]

Mitchell River group

Comprised mainly of men, from a vast catchment area in far north Queensland, the Mitchell River (MR) group's aim was to help manage the Mitchell River Watershed for the benefit of future generations and to prevent the type of environmental degradation witnessed in other great rivers in that State.[12] As the only voluntary policy forum for catchment management in Queensland at the time, the Mitchell River group were innovative and forward thinking.

It started at a meeting organised by Vic of the Kowanyama Aboriginal Land and Natural Resource Management Office (KALNRMO) in June 1990 at Kowanyama. All the major regional land management agencies attended with good representation from Federal as well as State government officers. There was widespread concern among graziers and fishermen, indigenous and non-indigenous about the condition of the Mitchell River and for watershed management generally. Arising from the conference was initial agreement on the vision, goals and objectives for the group (see Box 3.5) as well as a recommended membership list.

Box 3.5 The development of the Mitchell River group

The vision

To co-manage the Mitchell River catchment using an integrated approach to sustain the natural resources and maintain and improve quality and diversity of life.

Goals

- Prevent further degradation.
- Restore degraded resources.
- Ensure use of resources within their capability.
- Minimise adverse effects of resource use.
- Ensure self maintaining populations of native species.
- Ensure appropriate planning and management.
- Preserve cultural heritage.

Recommendations

That a steering committee be formed with the following terms of reference:

- To evaluate existing processes of integrated catchment management with a view to developing and recommending an appropriate management process for the Mitchell River catchment.
- To investigate the feasibility of further developing the process to cover the Cape York Peninsula.

Proceedings of June 1990 conference published by KALNRMO.

Men comprised the majority of the membership of the Mitchell River group. Garry is a retired shire councillor and grazier in his mid-sixties, who had moved to town and passed his cattle property over to his son to manage. He is very pro-development and claims that this country needs more jobs, more population and better services. Like others he joined the group to make sure that the "greenies" did not monopolise the group's agenda. He has a secondary school education, enjoys fishing, and doesn't get to church as often as he would like to. Josephine has been involved in conservative National Party politics for 30 years, she owns and manages an eco-tourism business, regularly attends church and takes a very "green" view of both environmental management and Aboriginal self-determination. Garry and Josephine do not get along and have been at the forefront of a continuing conflict in the group about development and the environment. Both believe that the group has prompted them to think and do more about environmental stewardship on their own land.

Right from the start, the Mitchell River group discussed management of the Watershed cooperatively with government, catchment users and others interested in the river system. Having become familiar with cooperative management (co-management) in the Pacific North West of the United States through exchange visits and videos from the Squaxin Island Tribal Council, co-management has always been an operating principle of the group.

The Mitchell River group helped raise community awareness about the health of the river - both as an ecosystem, a culturally significant site and as a source of income for professional fishers, tourism operators, miners and graziers. Despite the huge distances involved and the great divergence of interests, members of the group have stayed together despite some quite serious ideological conflicts. Perhaps their most important achievement has been maintaining their strength in diversity, and their ability to combine cultural, environmental and economic goals.

Lastly, the Mitchell River group has had by far the hardest task in developing a sense of community across a catchment which spans more than 75,000 square kilometres, 200 years or more of European and Asian history and tens of thousands of years of Aboriginal culture and history. The Mitchell River group has finished a program of short reports for their catchment in an effort to build a database of catchment characteristics, which draws upon both the socio-economic land use and biophysical nature of the region. These have contributed to a discussion paper that in turn has been used to flesh out a strategic plan for the watershed.[13] It includes position statements on the history of the catchment and the state of hydrology, vegetation, groundwater, biodiversity, climate and landform. It also summarises key issues facing the Mitchell and establishes a vision for the future based on cooperative resource management, education, information and communication. The Mitchell River Watershed Management Plan has been endorsed by the Management Group despite a deep-seated scepticism among some of its constituencies about the motivations and values of other groups within the catchment. This in itself is a major achievement for the Mitchell River group. Project work has begun in a number of areas within the catchment. These include weed management, ecological assessment of stock routes in the head waters and landcare work in the Cattle Creek sub-catchment to halt and prevent rising ground water on horticultural and agricultural land.

Despite these initiatives (among others not listed here), the involvement and commitment of an energetic and well informed local citizenry eludes the group. This is partially due to a lack of profile for the group, but mainly a result of a sparse population spread too thinly across a landscape with no concentrated urban centres. It may also be due to a legacy of independence, individualism and defiance left over from a colonising era and a frontier mentality, necessary in the early part of the 20th century but not so much of a virtue as we begin the 21st century.

Frequent changes in the facilitation structure and support for the group have also led to fragmentation of effort and a lack of institutional memory. This is not to decry the significant effort of individual facilitators. The founding facilitator left a very strong set of initiatives behind him. However, without this kind of full-time dedication to the group, subsequent facilitators were only able to achieve improvements on the edges of a plate that had been handed to them. The situation changed for the better with the appointment of a third full-time facilitator. She was appointed by the Mitchell River group itself rather than the Queensland Department of Natural Resources – they were responsible for dispensing integrated catchment on a less personal basis. The group has also made three changes to the role of chairperson and each one has lent a slightly different emphasis to its projects. From an initial interest in regional development initiatives there has been a subsequent concentration on Aboriginal reconciliation, grazing, weeds and pests, and environmental education. Notably, despite the changes in leadership, the group remains strong and viable. In essence, the mere existence of a group of people with such diverse skills, education, background, culture and experience makes the group's years of remaining in an establishment phase an amazing feat. The mere fact that group members are able to talk to each other at all is of considerable note. If anything, they have stronger horizontal ties than they do vertical ones. However, it is really too soon to assess the extent of the Mitchell River group's attachments to the local and wider community.

In summary, all three groups have made a distinct and important contribution to stewardship in their area. Each acted upon a perceived threat to their local environment, but the practices of group operation differed. As an illustration of their different profiles, see Table 3.1.

Table 3.1 Stewardship group comparison

PROFILE	WW	DL	MR
Membership	About 15	About 30	About 50
History	1991–1994	Started in 1989	Started in 1990
Focus of concern	Water quality	Dryland salinity	Watershed management
Government membership	No	No	Yes
Size of area	Approx. 500 km^2	Approx. 200 km^2	77,000 km^2
Membership fees	No	Yes – $15 pa	Yes $5.50 pa

Member involvement and motivation to join

For each stewardship group, there were some members who came to every event, and others who came occasionally. There were core members who did most of the work, and fringe members who were involved to a much lesser degree. Some members tended to spend a lot of time initially and then their involvement waned over the years as others joined, or with the employment of a group facilitator. On several occasions, members told me that their involvement has increased with new responsibilities and exciting new developments. As Mike said: *it could be my sole paid job and take up every hour of the day*. It is important to note that although Mike was the chairperson for his group, he was not paid by the group and held down a full-time position elsewhere in the community.

There are many reasons which motivate individuals to join a stewardship group, see Box 3.6.

Box 3.6 Joining a stewardship group: some motivating factors

Concern with a problem: care for country, desire for change, wish to prevent further degradation, frustration and dissatisfaction, interest in stewardship (landcare, water quality monitoring, catchment management).

Sense of community: district pride, social interaction, community cohesion, involvement and belonging, escape from routine, satisfaction and enjoyment, desire to participate, avoidance of loneliness.

Self-interest: access to financial grants, prestige, personal aspirations, prevent loss of personal advantage, NIMBY, issue affects an individual member personally.

Power of the group: good leadership, peer pressure, being asked to join, to have influence over others, for effective consultation, membership seems attractive or positive, group is well organised, group action is more effective than individual action.

Altruism: civic duty, commitment to an ideal or cause, ability to contribute, moral or ethical reasons, saving the world, being a good citizen, seeking a global perspective, making a local contribution.

Learning information/skills: exchanging ideas, gaining information, learning new skills, access to educational opportunities, gaining knowledge.

Source: interviews with group members

Canadian research into the motivations, satisfaction and rewards for volunteers working on the Ontario Breeding Bird Atlas confirms that volunteer involvement is strengthened when there is a range of motivating factors including a sense of purpose, duty, commitment, having time available, learning new skills, meeting new people, personal and global welfare or for enjoyment and fun.[14] Whether it is a focus on environmental stewardship more broadly or volunteer monitoring of water ways in particular, this report from the United States newsletter *Volunteer Monitor* sums it up:

> the heart of volunteer monitoring has always been the dedication of individuals to protecting water bodies that they live near and care about; about which they feel a sense of personal ownership. Ask volunteers why they got involved in monitoring and you'll often hear, "My grandchildren play in this creek," or "My family uses water from this lake as a drinking water supply," or "We don't want the place where we live to be ruined".[15]

Another Canadian stewardship scheme motivates its members in a variety of aesthetic, historical and practical ways. It is the stream daylighting movement, a network of local community groups who are "making plans to uncover and replenish long lost waterways".

One group member is reported as stating that "enthusiasm for daylighting initiatives 'comes from a desire to re-establish part of what has disappeared from our urban landscape'".[16] What motivates people to become involved in community groups is complex and cannot be simplified to issues of either the outcome or the process of involvement. Simple dichotomies based only on "good citizenship-altruism" or "material benefit – self-interest" models miss the subtleties and complexity of reasons for public involvement. Table 3.2 reflects the diversity of reasons which motivated members to join Water Watchers, the Downside Landcare or Mitchell River groups.

Table 3.2: Motivation for involvement in rank order

	Water Watchers		Downside Landcare		Mitchell River
1	Concern with a problem	1	Concern with a problem	1	Concern with a problem
2	Power of the group	2	To learn	2	Altruism
3	Involvement and belonging	3	To exchange ideas and information	3	To care for the country
4	Altruism	4	Self-interest	4	To have a say
5	Self-interest	5	Altruism	5	Power of the group
6	Social interaction	6	Social interaction		

Source: focus groups interviews

Motivation to join stewardship groups is stronger when individuals perceive a threat to the health and wellbeing of their families, their property or their community. For example, the threat of salinisation affecting the Downside community or poor water quality's effects on future generations were serious motivations for members of the Water Watchers and Mitchell River groups. Motivation to continue to be involved in group activity changes over time as individual circumstances change, and with the construction of ongoing relationships with other members. There are a number of factors which group members told me made it easier to be involved with environmental groups, see Box 3.7.

Box 3.7 What makes it easy to be involved

- Knowing, liking and fitting in with the group.
- Meeting and working locally.
- Having identified group goals, funding, meetings.
- Having time available for a worthwhile project.
- Having supportive government institutions.
- Having good leadership and coordination.
- Gaining and maintaining skills and knowledge.
- Satisfaction of achieving results.

Source: interviews with group members

Disincentives and restrictions

There are also a number of important factors which limit members' involvement with stewardship groups (Box 3.8). Not surprisingly the most important factor which restricts involvement in community activities, is lack of time. Where a group member gives time to stewardship, it is taken from work and other commitments (including farm work, housework, business and paid employment) as well as from home and family commitments (including personal hobbies and leisure time). Several group members told me that time spent in one group would adversely affect time spent in other community groups. This over-commitment represents a typical picture of restrictions to voluntarism, but also raises an important point concerning the role of environmental stewardship groups in Australia. Referring back to earlier comments, are "we" expecting "them" to do too much?

Box 3.8 What makes it hard to be involved

- Taking time off work.
- Family commitments.
- Distance from and location of meetings.
- Lack of special knowledge or skills.
- Lack of money for travel.
- Tired of going to meetings.

Source: interviews with group members

Several landcare group members thought that they had limited knowledge or understanding of scientific and technical matters. John said:

> I don't really know a lot about all that [landcare] stuff, but I help out a lot. I get most of my knowledge from Billie's newsletter – guest speakers bring pamphlets but I don't chase up the Department of Agriculture. I'm not into it as much as Mike growing all his trees ...

Inaccessible language is a big problem for volunteer stewards and is raised in more detail in Chapter 6. Restrictions reported by members of Water Watchers included such basic infrastructure as lack of a telephone, lack of access to transport and lack of childcare facilities. These were very significant for members of a peri-urban stewardship group and raise the issue of whether more people would have become involved with Water Watchers if only they had access to a telephone, a car, childcare services or other basic services. For the Mitchell River group, factors restricting involvement in the group were more likely to be travel and distance related, given the geographic extent of that group's operation.

Given this range of disincentives and restrictions for group members, it was interesting to note that government officers associated with each group gave very different explanations of what makes it hard for individual members to volunteer. Timing, structure and frequency of meetings as well as cliquishness, leadership problems, personality conflicts and discomfort with group process were all mentioned as disincentives by government employees. In other words, the structural impediments (such as time and resources) were not mentioned as frequently as in-group impediments of the kind mentioned. This perception may simply be due to government officers' failure to acknowledge the considerable time and effort put in by community members. Or, perhaps government officers are overly preoccupied with group process since participatory processes have been recently added to their duty statements. Until recently, these officers believed it was their job (and their job alone) to assist community groups with the more technical aspects of environmental problems. Now that groups are becoming more active without help from these traditional technical authorities and that departmental policy favours participatory processes, some extension scientists and government bureaucrats are feeling uncomfortable with the social processes and facilitation skills now required to manage group dynamics. This in turn may be projected onto the newly forming groups and subsequently be raised as a disincentive to join a stewardship initiative or activity managed by a local group.

The roles of stewardship groups

Stewardship groups take on a wide range of environmental management roles. Members of Water Watchers, Downside Landcare and the Mitchell River groups were asked to consider their stewardship roles in detail. Two categories of roles emerged. *Major* roles were those identified more frequently by group members and describe *what* groups do in terms of stewardship. *Minor* roles emphasise *how* the action takes place. See Boxes 3.9 and 3.10.

Box 3.9 Stewardship groups: major roles

Change agent: awareness and recognition of the problem, identifying an issue, defining goals and activities, problem-solving, being a catalyst for change.

Local educator: providing a means for schools' involvement, environmental education, source of knowledge about local communities.

Information/skills banker: information exchange, skills exchange, information source, focal point for collection of expertise, information collection and dissemination.

Advocate: vehicle for expression of community needs, wants, reflection of ideals and communication of those to government, lobbying.

Administrator: being cost-efficient, raising money, managing projects, facilitating, consultant to other groups, providing a voluntary labour source.

Restorer: active environmental maintenance and protection, practical hands-on work, taking environmental responsibility, future environmental protection.

Box 3.10 Stewardship groups: minor roles

Community actor: working together for the common good, toward common goals, implies that collective action is greater than individual action.

Partner: networking, mediating between agencies and individuals, contact-sharing, acting in a liaison capacity.

Watchdog: alerting others to potential or existing environmental problems, being responsive, acting fast.

Social influence: peer pressure, influencing others within the group, setting up in-group norms and behaviour.

Facilitator: encouraging involvement and participation within the wider community, acting as a catalyst for involvement, providing an example.

Acting as a change agent was the most important stewardship role identified by members of each group. It includes becoming aware of or recognising the problem or issue to be addressed by the group, as well as setting goals and strategies to manage it. In the United States, this desire for local change is similarly important to volunteers in their efforts toward environmental stewardship. There are many published descriptions of local practical restoration efforts, but perhaps none so positive as the effort to rehabilitate a salt marsh on Rhode Island by digging a channel between two creeks.

> The typical marsh restoration project requires bulldozers, trained operators, and weeks of work. But the small scale of the Little Mussachuck project gave volunteers the rare opportunity to perform all the work by hand. A crew of 70 volunteers ranging from high school students to retired engineers turned out in April 1998, equipped with shovels and buckets. They took just four days to dig a 300-foot channel to connect the two creeks. The route of the channel had been planned out ahead, based on a survey of the natural contours of the marsh, and as the crew dug, Save The Bay staff used a laser level to ensure proper depth and width.[17]

This emphasis on "do-it-yourself" practical projects is typical of stewardship groups, but so is serving the needs of group members themselves. Members of all three stewardship groups I visited spoke frequently of the importance of maintaining the cohesive function and power of the group. Government officers did not identify these social functions to the same degree, and were more likely to suggest that community groups have a role in creating partnerships with government. Several officers even suggested that community groups should extend the work of government by providing voluntary environmental watchdog services. Notably, one government official reported that community groups should have no role in environmental management at all:

> They're spoilers, not pro-development and a waste of time for government ... They use media and others forms [of publicity] to bring undue pressure on development without full knowledge or technical expertise. (Geoff, a Mitchell River government officer)

Skills and personal qualities of group members

So what is it about the members of community groups that qualifies them to undertake environmental stewardship? I asked: "What skills, experience and personal attributes of individual group members is necessary for a community group to be effective?" Some said that groups need local, practical, community-based or experiential knowledge (for example Daisy of Downside said you need, "life experiences"). Others mentioned "outside experience" or knowledge gained from government, universities or other educational institutions. Members frequently told me that there was a need for a diversity of experiences within a group. They said that it did not matter whether any one individual had certain experiences or not, as long as there was a range of collective experiences, skills and personal qualities to draw upon. Mel of Downside went as far as to say that "anyone can help – it's not essential to have any special skills or qualities". Most people however told me of the need for specific types of skills and personal qualities. I categorised these and present them in Box 3.11 and 3.12, respectively.

Box 3.11 Stewardship group: skills base

Negotiation skills include mediation, conciliation, conflict resolution, networking and liaison skills.

Group process skills include decision-making, priority and goal setting, long-term planning and time-keeping skills.

Communication skills include both oral and written skills such as letter-writing, public speaking, listening skills and public relations skills.

Organisational skills include secretarial skills, management skills, book keeping skills, financial skills, fundraising skills and meeting skills such as "chairpersonship".

Leadership skills include delegation, motivational skills, facilitation and coordination skills.

Research and analytical skills include specific technical abilities, problem-solving ability, logic skills and research skills in terms of where to access information and how to contact people or mobilise resources.

Communication skills were very important to members of all three stewardship groups. Interestingly, group processes which include such skills as decision-making, goal-setting, planning and time-management were cited less frequently by group members and yet are given the most attention by academics and government advisers in encouraging groups to become established. To a lesser extent, organisational skills, leadership skills and research and analytical skills were also more highly valued by government officers. The reasons for this are not yet obvious, but will be discussed in Chapter 5. Group members emphasised the necessity of good interpersonal qualities for effective group action.

Box 3.12 Personal attributes of stewardship group members

Commitment and persistence relates to personal qualities which included discipline, stamina, perseverance, determination, "follow-through", stalwartness and continuity.

Creativity and vision relates to personal qualities of inspiration, open-mindedness, initiative, inventiveness, desire for change, holism, and ability for forward thinking and being able to see a broader perspective.

An historical perspective relates to being able to identify local problems, understand local conditions and environments, make contacts and maintain networks with local people.

Outcome-orientation relates to being able to achieve, be competent, responsive, quick witted, maintain a work ethic and to get results.

Motivation and positivity implied concepts such as enthusiasm, charisma and presence, a willingness to learn, interest, a positive spirit, a sense of fun, being encouraging of others and having an outgoing personality.

Community-orientation relates to a willingness to share, being community minded, having common values and goals and being accepted into the community.

Nurturing relates to an ability to get on with people, friendliness, affability, patience, warmth, cooperation, being supportive of other group members and of the leadership, and being able to take direction.

Consideration included diplomacy, sensitivity, tact, empathy, trust, tolerance, an ability to compromise, to recognise others' skills, to be supportive of others' involvement, to be respectful of difference and of others' views, and to be able to bring people together.

Downside Landcare group members listed personal qualities such as commitment and persistence, motivation and positivity, community orientation and nurturing abilities as essential ingredients of individual members in order for the group to be successful. I believe this is firmly linked to their strong sense of community and sense of place, discussed further in Chapter 6. On the other hand, they were not as outcome-oriented as members of the other two groups. Perhaps they were reinforcing the belief that in order for groups to be effective, individual members have to give up their rugged individualist stance, and trade off their capacity to get the job done, to demonstrate an ability to "act fast and work hard", with more group-oriented personal qualities. Similarly, members of the Mitchell River group emphasised the need for tolerance, respect and cooperation. Given the wide diversity of their membership base, cultural heritage and educational status, it is not surprising that they emphasised the importance of diversity.

This profile of community groups and their members helps to answer some questions, such as who belongs, what stewardship roles do they play and what makes it easy or difficult for members to join. Chapter Four provides more detail about how groups actually work.

Endnotes

1 Water Watchers formed before the Australian national Water Watch program started and were not a subset of the national program. Their data was therefore not part of the national program although they did send it to the Water Authority of Western Australia.

2 See Brunson's and Landres' chapters in Knight and Landres 1998 *Stewardship across boundaries* for a good overview of the social, political and ecological dimensions of maintaining boundaries for natural resource management.

3 Furthermore to protect anonymity, when giving a profile of membership in each group, I have described members in composite terms, they are not actual descriptions of any individual.

4 Gardner 1991a and b.

5 That definition was developed by the Society for Ecological Resoration Policy Working Group in 1996 after almost a year of consultation and deliberation. It is taken from the SER web site: *http://ser.org/definitions.html*.

6 Adapted from Carley and Christie 1992.

7 Lerner 1994.

8 Rush and Associates 1992. While there has been a lot of interim research as to why individuals belong to groups, there has not been any further research indicating why groups form.

9 For more details, see Carr 1993a.

10 For more details on Water Watchers, see Carr 1992.

11 Star 2000: 69.

12 For further details on the Mitchell River group, see Carr 1993b.

13 MRWMg 1999.

14 Ordubegian 1993.

15 Taken from Anon, "A Profile of Volunteer Monitoring" The Volunteer Monitor, Vol. 6, No. 1, Spring 1994 published electronically: *http://www.epa.gov/volunteer/spring94/ppresf07.htm*.

16 Botelho 1999.

17 Ferguson wrote this description for The Volunteer Monitor Vol. 11, No. 1, Spring 1999: *http://www.epa.gov/volunteer/spring99*.

4

Stewardship groups: in practice

This chapter moves beyond looking at a profile of stewardship groups and begins to investigate how they function – the practices involved in how groups really work. In order to do that, it is important to consider groups' goals and objectives, meetings and decision-making structures, leadership and facilitation techniques, education and training provisions and their consequent "products" or outcomes. Although this chapter concentrates on the experience of Water Watchers, Downside Landcare and the Mitchell River groups, attention is drawn to the literature about small group function where appropriate.[1]

Goals

The establishment of clear, executable goals and objectives has become widely recognised as one of the factors which contributes most to the success of community groups, regardless of their endeavour. The way in which a group starts with or without a common purpose or long-term set of goals can fundamentally affect group process and function. It is important therefore to consider how a group engages in goal or objective-setting through the use of minutes, written memo's, notes, etc workshop sessions or futures building.

It is interesting to note similarities between the goals of the Water Watchers, Downside Landcare and the Mitchell River groups (see Box 4.1). Water Watchers group members were very clear about what they planned to achieve – water quality monitoring. This was perceived by most members as their only goal, as the Coordinator for the group, Louise explained: "The aims were written up eventually. In our minds we always had them, but they weren't formulated until I had to give some talks". Richard, the government facilitator for this group spoke enthusiastically about Water Watchers' goal setting process:

> Common goals were worked out after what we wanted to achieve – fairly early on. It was on the whiteboard, so everyone knew what they had to do. I did like the way everyone was listened to and participated.

Box 4.1 Stewardship group goals

Water Watchers

- To collect data/information on phosphorus loadings and salinity in water quality within [the coastal catchment of] the Serpentine Jarrahdale Shire.
- To pinpoint specific water quality issues according to land use.
- To find a pattern or correlation between land use and water quality in the Peel Harvey inlet.
- To provide a base standard of water quality data for future planning.
- To raise community awareness of the health of the waterways.
- To bring a variety of people together to work alongside each other.
- To demonstrate that it wasn't just farmers at fault.
- To develop a water quality monitoring system for the community.
- To put the information collected into a form that could be used by others.
- To disseminate the information.

Downside Landcare

- To solve the problems [acidity, salinity] of the area.
- To learn about these problems and why they're happening.
- To make unproductive land stable or productive.
- To see and understand the problems.
- To work with a group to achieve change.
- To get people involved.
- To create awareness in others.

Mitchell River

- To encourage sustainable development.
- To identify issues and stakeholders in the watershed.
- To look after the watershed.
- To establish cooperation among the users.
- To identify resources in the catchment and minimise [adverse] impacts.
- To recognise and respect cultural heritage.
- To educate existing stakeholders about sustainable land use practices.
- To get everyone together talking.
- To prevent further degradation.
- To recognise and promote the bottom-up [community control] approach.
- To save the environment for future generations.
- To make a public plan for the catchment.
- To save fisheries regarding the declining fish stocks.

In Downside, group members were divided between those who thought there were goals initially, and those who were sure that there weren't any initial goals:

> Yes we did [have goals]. We had to have goals to get government assistance to form a landcare group - like fighting land degradation and having it set up as a community project. (Keith, Downside Landcare group member)
>
> No - [we didn't have goals] we took the attitude of you've got to crawl before you walk ... Let's continue "seeing is believing". Let's make it happen rather than writing it all down. (Bill, Downside Landcare group member)

Mitchell River group members defined their goals and objectives at an initial conference when the Watershed group was formed. However, again there was some confusion and little real ownership of these goals:

> I can't remember exactly, but we did have 8-10 goals and we discussed them and all agreed to them. (Phil, Mitchell River group member)
>
> Out of the conference some general goals and terms of reference for the Mitchell River group were set. What was lacking for a little while was smaller goals and strategies on what to do. (Vic, Mitchell River group member)

Although stewardship groups often have formalised goals and a written vision statement, members have little attachment to goals that they do not participate in setting, or that are not written in simple language. While successful groups need formal goals and objectives by which to measure their progress, the emphasis should be upon formulating and acting upon practical goals which are clearly defined, specific, achievable, well understood and agreed to by consensus. As was the case for the Water Watchers and Downside Landcare groups, perhaps these goals should not be articulated "up-front", but left until one project has been completed and members have a better understanding as to what they wish to achieve next.

Goals and objectives are clearly translated into everyday language about local projects in the minds of many stewardship group members. Contrary to traditional, strategic or comprehensive planning theory which states that people generally make goals in advance based upon a SWOT analysis (Strengths, Weaknesses, Opportunities and Threats) and then act to carry them through, stewardship group members gain much through goals which are formed and clarified through practical action and experience, rather than in the abstract at the beginning of a project. Many group planning processes emulate cyclical theories of action learning whereby action and reflection are continuous loops, defining and refining goals. Community-based planning is iterative; groups visit all the stages described by traditional planning theory, but may start at different stages in the planning cycle. Also, since they don't always have obligations to particular planning "outcomes" set by bureaucracies according to time-lines and flow charts, they move at their own rate, backwards as well as forwards, repeating different stages as and when necessary.

Meetings and decision-making structures

Community groups vary considerably in their structure and style of decision-making. Some meet whenever necessary, often at short notice, to discuss their stewardship strategies and to distribute information. Decisions made at these meetings often concern the very practical nature of particular projects. In the case of the Water Watchers group, intra-group communication took place via a telephone tree which was informally and later formally established to keep everyone up to date. The facilitator was the central hub of the telephone tree system, directing and redirecting information to group members. The telephone tree had the effect of involving more people in decision-making initially, but at the same time, taking longer to arrive at a mutually agreeable course of action, and eventually it disintegrated.

Other groups are more structured in their decision-making approach, incorporating public meetings with the wider community and perhaps an executive committee to "keep the pot boiling" and handle decisions that arise on a month-to-month basis. In the case of the Downside Landcare group, an executive committee comprised office bearers and several general members. Before the group had an official coordinator, many of the day-to-day decisions were made by the chairperson and secretary of the group. Decisions about the direction of the group were made at open community meetings and in liaison with the "executive" either face-to-face or by telephone. In accordance with the power vested in the executive committee (and especially the chairperson) by the rest of the landcare group, members were generally happy to have day-to-day decisions made for them about the detailed running of the group. In a very different model to that of Water Watchers, Downside Landcare group members subscribed to the practice of electing members to committee positions and then letting them get on with the job.

As with the Downside Landcare group, most Mitchell River group members were happy to have day-to-day decisions made by the chairperson who was highly respected in the area. However, difficult questions arose concerning decision-making powers when new members rejected decisions made by a former chairperson and wanted a more formal process. Conflict erupted when some original members insisted that group management practices should not change. The difference of opinions centred around the extent to which the Mitchell River group was a community-based group allowing decisions to emerge from a consensual decision-making process. This conflict was exacerbated when several new members wanted to introduce formal voting procedures and expressed the view that not every "Tom, Dick and Harry" should attend group meetings because the process and structure of the meetings would be too long. The problem of some members being able to "stack" meetings when voting on important issues was also brought up. This debate raised the issue of who should be a member of the Mitchell River group and what constitution the group should adopt.

Meeting styles and decision-making structures play an important part in the democratic processes operating in groups. The relaxed casual style which worked for Water Watchers would not have been appropriate for the Downside group. As noted above, stewardship groups vary in their approach to participative democracy. This has a lot to do with local culture and geographic context. The important point to note for facilitators, councillors, rural liaison officers, and other government officers, is not to expect conformity of democratic processes. The number, frequency, style and timing of meetings has to be locally determined and set in accordance with the type of community in which a group is located. While that may seem obvious and common sensical, what is less often acknowledged by volunteer coordinators, chairpersons and secretaries is the need to put decision-making practices on the agenda right from the outset of a group's establishment.

In order for groups to evolve and develop as effective stewardship agents, there must be some degree of healthy self-reflection and the promotion of open, direct communication about the group process itself. Too often group members construct their roles as purely content-focused and neglect the "house-keeping" processes which ultimately determine the cohesion and strength of a group's social relations. For the Downside Landcare and Mitchell River groups, this lack of self-reflection led to an assumption about the formal way that group meetings should be conducted. "Roberts Rules" and "due process" prevailed with the chairperson and facilitator sitting at the front of the room and all business being conducted through the chairperson. Members believed that the group would not be legitimate or official if meetings were not conducted properly:

> Oh yes we keep minutes, there is a formal part of the meeting. We're just about to adopt a constitution and become incorporated. (Glenn, Mitchell River group member)

> [Meetings are] very formal because we had committee nominations. Minutes were kept so that we could prove we were a viable group to gain funds and show that we were serious with it. (Rhiannon, Downside Landcare group member)

The level of formality is often a big issue for a stewardship group trying to "prove" their legitimacy. However, this perceived need to be accountable (to government or non-government funding agencies) can lead to issues of non-involvement, disempowerment, lack of consultation and lack of representation for a minority who believe they are unable to affect group decisions.

Representation

Representation is a difficult issue for many stewardship groups. Some interpret it to mean geographically representative of an area using land-use types or topographic features (north or south of the railway, up or down a valley). There are others who interpret it to mean representative of the population of an area, as in employment type, age and community status. Brian of the Mitchell River group said:

> [We] are representative of the land users and people types and attitudes, but not representative of all the people who live in the catchment.

Stewardship group members commonly interpret representation to mean including a diversity of interests or stakeholders on a committee in a specific environmental context. Landcare group members define their constituency generally as farmers, or landholders, within a defined geographic area. Increasingly, there is a demand by so-called "outsiders" to have their issues and concerns addressed. These are not just geographic outsiders – people who live on the fringe of a landcare area or in town – but outsiders in the sense that they are seldom seen to have the legitimate interests of people "on the inside". Some stewardship groups are attempting to involve a broader range of interests and people in their proposed activity. The Downside Landcare group are a good example. They involved "hobby-farmers", retired school teachers, members of a local tree-planting group, school-children and anyone else who showed an interest.

I believe that a more creative interpretation of stakeholder representation is critical for environmental stewardship groups. Those who hold a stake in environmental issues could be interpreted very loosely: future generations, traditional custodians of an area, consumers of food and fibre, artists, overseas markets … the non-human world. More conservative representation is the norm in stewardship groups. People who consider they have a right to be involved often include local landholders, government officials and elected representatives. Simultaneously (and often not consciously) those who consider they have an environmental responsibility but lack power or status believe they are excluded. Members of the Water Watchers group felt strongly about ineffective representation and exclusion:

> We are the only ones who don't have property in the district. Local fishermen in Mandurah are not represented either, but they have a vital interest. (Catriona, Water Watchers group member)
>
> [Membership is] very cliquey, too many big farmers and councillors [It's] too elitist. They all have handles to their brooms. (Janine, Water Watchers group member)

Members of the Mitchell River group wanted to define their representation more broadly. Brian wanted to add more river users and fishing groups to the group. Stuart wanted to see representation from "some of the agricultural interests around the tablelands – [there are] none from sugar or tobacco or small cropping sectors and most of these are near the headwaters [of the Mitchell River]". Barny wanted to "see extra representatives from Kowanyama, from the Aboriginal people themselves". Other members agreed believing that while democratic representation may be very difficult to achieve, it is nevertheless a worthy pursuit. In particular it was felt that Aboriginal groups from other parts of the catchment should become involved in the Mitchell River group. For all three groups, more representation and more effective representation are critical factors in the longevity of stewardship groups. Another such issue is that of leadership and facilitation.

Leadership and facilitation

Almost all stewardship groups have a clearly recognised set of leaders. These voluntary coordinators often take responsibility for enlisting potential participants, organising meetings, writing funding submissions and liaising with the media or general public. In new groups they are usually, but not always, the founders of a group and are widely perceived to be the primary "movers and shakers". For Water Watchers (Louise and Jenny), Downside Landcare (Mike and Ellen) and the Mitchell River group (Vic), one or more people could clearly be identified in this role and were generally supported by the rest of their group.

In Australia, many stewardship groups have access to the services of a government-funded facilitator. This was the case for the Water Watchers, Downside Landcare and the Mitchell River groups. Often the facilitator's position is created and advertised by a stewardship group or, by several neighbouring groups. There is however a great deal of variation between groups in the terms of a facilitator's appointment. Some are partly funded by the group itself, which implies a different kind of working relationship between the group and its facilitator. Other positions, although fully funded by government, may be locally accountable to, and appointed by, a stewardship group. The nature of both voluntary leaders' and paid facilitators' relationships with stewardship groups is critical to group process and may largely determine the level of success in achieving specific stewardship goals and strategies.

In the case of the Downside Landcare group, the key people responsible for the group included the chairperson, facilitator, secretary, treasurer, publicity officer and several other executive committee representatives. Members spoke of the inaugural role of Mike as chairperson and Ellen as secretary in setting out to motivate local landholders to become involved in the landcare group. As the group matured however, the situation changed. Cathy was employed as facilitator by the group with funding from the National Landcare Program via the New South Wales

Department of Conservation and Land Management. She lived in the district and owned land there which had been passed down through several generations. She was consistently lauded by group members as someone who took over most of the administrative load, assisted with the newsletter and managed liaison with government departments.

A different model of leadership was reported by the Water Watchers members. Their facilitator, Richard, was paid by the Western Australian Department of Agriculture and based at the Community Catchment Centre. He also spent a considerable amount of time talking with, and assisting, group members. However, his role as a leader was compromised. He preferred to describe himself as a member of the group rather than someone paid to facilitate. Yet, when an opportunity arose, he left the group for a position in a different State. Other group members did not perceive him to truly belong in the sense that he did not live in the area and did not have a long history of attachment to the community. This issue is raised again in Chapter 5.

Problems of transference from one leader to the next are also very common with voluntary coordinators or chairpersons. Mike, the Downside Landcare group's chairperson was concerned that he was "running the show". He wanted to attract other potential leaders to the position, in order to hand over the leadership reins to someone else. He knew that part of good leadership was letting go:

> I've had too much involvement [in the group] according to my family. Although I'd be keen to put more in, I have to restrict myself a bit ... I'm doing a lot more now than my family would like me to do – to show how well it can work – then others will get involved later on.

Although he was co-founder of the group, he had not always been the chairperson. For a one-year period Mike stepped down from this position. Several group members remembered an incident at a committee meeting held to elect new office bearers. When no-one new was prepared to take on the chairpersonship, Mike was re-elected by default. This phenomenon is not peculiar to the Downside Landcare group. Reflecting commonly held beliefs on this subject, Samuel employed by the NSW Department of Agriculture told me that: "leadership skills are the main factor in self-perpetuation of groups – and are also a major stumbling block". Strategies for encouraging voluntary group leadership may include "increasing the pool of potential leaders, augmenting the personal resources of leaders and encouraging entrepreneurs".[2]

The reticence of other community members to step into the leadership position in Downside (as with other landcare groups) has to do with:

- the perceived lack of competence and charisma in other landcare group members; and
- an unwillingness to shake the status quo according to unwritten rule of conduct associated with knocking the powerful and existing landcare leaders.

This last point is worthy of expansion. Australian rural communities are riddled with unwritten rules and laws of social conduct which promote division among rural society, and have a distinctive influence upon group leadership. Much of this activity has to do with being an "insider" or an "outsider" to rural communities. Insiders in the Downside Landcare group belong to the tennis club, go to church, use Ultra High Frequency (UHF) radio and have an identified family history associated with generations on the land. Outsiders do not belong to these groups, have no identified social history and generally do not mix with others in the local community. Being on the outside does not necessarily mean being a newcomer to the district, although there is often a "waiting period" necessary for social acceptability. Some landholders in this category may have been living in the community for 20 years or more and still do not associate with others in the district. In other words, small rural communities do not always live up to the romantic and rustic ideals of what is projected in the media as one big happy family.[3]

Leadership in these circumstances is often preserved for the more established, well respected members of the "in" community. This process is referred to as one of "clique elitism";[4] referring to ways in which a small group or individual has power over resources, decision-making and the flow of information in rural communities. The implication of this form of control is a loss of public participation in landcare decisions. High-status groups, such as the executive committee on stewardship groups, sometimes use their power and ability to articulate their needs to the detriment of low status groups, or "outsiders" in rural communities, thereby diminishing the pluralist nature of participation in environmental stewardship.[5] Although this overt model of elitism was not operating in the Water Watchers, Downside Landcare or the

Mitchell River groups described here, it is important to recognise that stewardship groups in general and landcare groups in particular, are not just another local organisation. They are part of a broader socio-political phenomenon with attendant issues of legitimacy, kudos and hierarchy.

All stewardship groups must pay attention to issues of exclusion, elitism, transference and participative democracy in relation to leadership and facilitation. As with Rotary, Lions or Apex Clubs, attempts to "groom" another group member for the chairperson's position over the course of a year met with success in Downside. Like Mike did originally, current landcare group leaders rely on support from the executive committee and actively promote involvement of all group members. These factors are an indication that a group is well on its way to maturity.

Conflict and the group process

Conflict is inevitable in most stewardship groups. Importantly, however, conflict is not necessarily an indication of poor management within a group. On the contrary, it may signal the need for change, or indicate that the issues being confronted are inherently difficult ones. For most stewardship groups perseverance and loyalty to the group process will pay off, even when that process involves considerable conflict. Effective management of conflict reflects a healthy group in which diversity of opinion is welcomed. Even though conflict was present in the Water Watchers, Downside Landcare and Mitchell River groups, the effect of these conflicts on group process and outcome varied considerably.

In Water Watchers, conflict surfaced concerning who belonged and who was perceived as excluded from the group. Although funded originally from the Office of Women's Interests as an environmental project for women, at no stage were men ever excluded. Despite this, Robbo and Daniel expressed concern that it had ever been perceived as a women's organisation which led to some divisiveness, a dormant conflict which never quite erupted. Another level at which conflict occurred in the group was in the breakdown of communication between group members, made explicit with the demise of a telephone tree system for passing on information. Several members expressed disappointment about not being told of a certain event or not being informed about the latest occurrences. This lack of group cohesion and communication was exacerbated during the change over of government facilitators between monitoring seasons. The result of one meeting called in a hurry was that several people were not informed of the date and so were unable to attend. Over the next few months before a new facilitator was instated, minutes were kept by another member and distributed only to those who attended the interim meetings. At the same time, membership changed and expanded and founding members took less of a leadership role. By the time those who were part of the original membership list were invited to a group meeting with the new facilitator, (who had a mailing database on record), three interim meetings had been held. The rapid change shocked some original members. One said: "there was almost a completely new group last night". Three months later, this person no longer believed he was part of the group, its goals or its future.

> I think that in the December meeting there were certain people who were excluded on purpose. There was a lot of politicking going on. (Buster, Water Watchers group member)

Conflict among Downside Landcare members led to a difficult situation early in the development of the group which culminated in the chairperson standing down from his position. The story is a familiar one in rural Australia. An issue of farm size - whether landholders should be allowed to subdivide their blocks or not - split the whole community. Some of the group's members supported sub-division and others did not. It was perceived by some as a contest between the haves (not wishing good quality agricultural land to be subdivided for either hobby-farmers or housing blocks) and the have-nots (who needed the money to continue to farm on a reduced acreage). The group's chairperson Mike perceived the conflict was affecting his integrity and fairness as group leader, so he resigned, explaining: "I didn't want to be seen to be a key person [supporting anti-subdivision]". Given Mike's honesty and the transparent way in which he handled a potentially very divisive situation, that conflict did not leave any major scars on the group. Subsequently, Mike was reinstated by landcare group members. Although the subdivision conflict was a low point for the group, most members believed that handling conflict successfully was a positive experience for the group as a whole.

Without conflict in the Mitchell River group, undoubtedly the group would not have developed in the way that it did. Conflicts within the Mitchell River group were pervasive and profound, (not least because of the makeup of the group itself) and remain an ongoing theme in its evolution. In perhaps the most unique of Australia's catchment management committees, there was a mix of people who are traditionally perceived as not being the best of friends. At the time I spoke to them, the group was comprised of environmentalists, graziers, Aborigines, miners, tourist operators and professional fishers. There were also a lot of government departments represented on the group who were, at times, in competition among themselves (see Chapter 5 for more detail on this). Potential group conflict also existed within the context of external threats imposed by such political debates as reconciliation with indigenous Australians and economic conditions such as the collapse in the local tobacco industry. As a background comment to illustrate the conflicts inherent in the reconciliation theme, one of the government advisers, Blane, told me that a grazier's wife would give her husband's second-hand clothes to Aboriginal men and women working on the station. While he reported the donation as a charitable act, in the same sentence he told me that this grazier's mother-in-law would still refer to these indigenous workers as boys and girls, even if they were older and well respected in the community. Blane said "there are two levels of meaning here. On the surface it's OK, but deeper down, there's racism".

Differences of opinion arose between representatives of different factions or industry groups. These however were limited to situations in which for example, the regular representative of the Cairns and Far North Environment Centre was absent and an alternative CAFNEC member, Jim, attended the bi-monthly group meeting. Being unfamiliar with existing group dynamics and other members' personalities, Jim upset established relationships and retarded consensus on, or negotiation of, difficult decisions on several occasions. From where I sat, the pushing of positions on behalf of certain organisations and industry groups certainly seemed to curtail discussion of the issues and distract members from looking for alternatives. An illustration of extreme conflict in the Mitchell River group serves to underscore the importance of conflict management.

Early in the group's establishment, a development issue threatened to divide the membership over the potential ecological and economic impact of a tourist facility. A local developer proposed building a Tropical Highland Residential Resort for elite international visitors. Part of this development was the establishment of a series of dams near the headwaters of the Mitchell River. Local government officials could see the benefits of spin-off employment and economic recovery for an area with high levels of unemployment. Gareth, an enthusiastic supporter told me that: "it is the best thing to happen to Mareeba since sliced bread". On the other hand, the Cairns and Far North Environment Centre were vehemently opposed to the development on ecological grounds. Jim, Gillian and several other group members wanted to prove there would be no ecological ill effects before the development went any further. They advocated undertaking an environmental impact study to monitor baseline changes in environmental attributes of the site. Brochures advertising the development maintained that: "The first Southedge Dam has, since 1988, been a statement of environmental impact far superior to any academic study which could have been devised".[6] At an open meeting between the developer and Mitchell River group members, I observed the core of an ideological difference so deep-seated it would be very hard to reconcile. The developer said: "environmental impact statements are three expensive words ... you have to break the eggs to make an omelette". The point he was making was premised on a firm belief that: "so often scientists are wrong ... where are all the do-ers? If we sit on top of the mountain and do nothing – who will pay the tucker bills? If you do nothing you get nothing back". The difference in values and opinions was unlikely to be resolved given that it was founded on two such completely different world-views.

That conflict escalated early in the establishment phase of the Mitchell River group when the chairperson at the time made a public statement decrying the bulldozing of trees into the creeks on the Southedge property. Some group members believed that the media release did not represent the views of the whole group. During and after this disagreement, members aligned themselves with one perspective or the other which served to divide the group further and pit hard-won friendships and liaisons against one another in a very destructive way. Others successfully defused the situation by maintaining and developing group cohesion and refusing to take sides. Fortunately, some ground was able to be made up between the two most vocal representatives of each "camp" and the conflict did not have any lasting effects, despite newcomers' attempts to re-ignite the issue. That situation illustrates how conflict can damage budding relationships in any community group. At the same time, it is indicative of the resilience and flexibility of stewardship groups who achieve a great deal by not getting locked into positions and maintaining clear communication about the issues.

Education and training

Within stewardship groups, investments in education, training and skills development can pay off even in the short term. Members of the Water Watchers, Downside Landcare and Mitchell River groups confirmed that learning occurred for them on a personal, within-group, community-wide and societal scale. Certainly at the personal and local end of that continuum, many people expressed a feeling of satisfaction with their opportunities for learning:

> I've learned heaps. I've been amazed and flattered by the Aboriginal people and the way they've treated me, with honour I've never experienced from any other group of people. I've learned how graziers interact and why miners have wronged them in the past. I've learned the codes of conduct between Aborigines, miners and graziers and that body language is as important as language. I've learned to chair a diverse group and am still learning how to run these meetings. (Brian, Mitchell River group member)

Much of this "education" took place on an informal level with co-workers and from a diverse range of on-the-job experiences. Downside Landcare group members spoke favourably about the skills they learned from other people both within the group and from the local community. Neville said: "the Soil Conservation and Department of Agriculture blokes - you learn ideas off them". Mike, the group's chairperson told me he'd developed "people skills, conciliatory skills, more tact". And Cathy, the Downside Landcare group facilitator told me at length about what she had learned:

> I've developed more confidence in speaking my mind; more realisation of my self-worth and the skills I have to offer; I've increased my knowledge on problems and solutions - specific to the Downside area and generally in the Wagga area. I have a better understanding of people, [I've] learned how to encourage people to use their skills better and have started learning how

to delegate. [I have] more knowledge about trees and the confidence to lime and directly drill pastures.

Formal opportunities for learning new skills were also offered in each of the stewardship groups. In Downside, workshops on whole farm planning were held and guest speakers invited to meetings to talk about landcare statistics in the area. Water Watchers provided many opportunities for formal education as members learned how to monitor water quality and calculate phosphorus loads. In the Mitchell River group, formal educational opportunities occurred less frequently, but the role of local ecological knowledge and its dissemination for non-Aboriginal group members is significant and is discussed further in Chapter 6.

It is important for stewardship groups to recognise the potential opportunities for promoting broader environmental awareness via formal training. This can be as simple as organising an on-farm walk, or it can be extended to include invitations to guest speakers at meetings and field days, establishing public displays, arranging workshops and slide-shows and using local newspaper, radio and television outlets. Different approaches to, and philosophies behind, environmental awareness and education are also discussed more fully in Chapter 6.

Outcomes and products

Although the goals of any specific stewardship group will be unique, there are several broad categories of outcomes, achievements or products which unite all stewardship groups:

1. public involvement and awareness;
2. stewardship action;
3. education and dissemination of information.

Rather than attempting to evaluate the Water Watchers, Downside Landcare or Mitchell River groups' achievements against these three criteria in any objective or independent manner, it is useful to consider group members' own perceptions and appraisals of their performance. Although these refer specifically to their individual experience and cannot be generalised across the broad range of community-based environmental groups, their comments are often insightful. Both the "good" and "bad" achievements are discussed by group members and are summarised in Tables 4.1, 4.2 and 4.3, corresponding to the three main categories noted above.

Table 4.1 Public involvement and awareness

Bad points	Good points
Water Watchers Parochialism. Not enough people resources. Our school was not involved. The school's problem with insurance [regarding taking children outside school premises].	Involvement of 50 kids, they were told the results, they had fun, educating their parents through hands-on involvement. We were making a positive contribution. We are more educated and connected [than before]. The continuation of water quality monitoring from sampling to analysis. Involvement with the British Petroleum (BP) awards. Radio interviews and television programs. The Serpentine Fair and Royal Show stands and leaflets published.
Downside Landcare Some [landholders] haven't joined and you can't make people join. Some properties may be devalued because of awareness of salinity. Is it our job to create awareness in others? Sign-posts don't tell you what it's about. It can drag on, some people get sick of doing everything. Busy life – you can't get involved. Costs money to be involved. Frustration. If you can't pay for fencing, you can't get involved. People are afraid of involvement. Not everybody wants to get elected to [committee] positions.	The Benalla bus trip was good for those who went and because of it the group formed. [We have an] active publicity officer, who keeps photographic records. The Newsletter keeps people informed – it creates enthusiasm and acts as a reminder of our achievements. School kids involved. Fun, chatter, comradeship. Good learning. It's interesting it motivates you to see others involved. You become more understanding and more tolerant of others' problems. Chemical companies were involved – free PR.
Mitchell River Facilitator has split alliances [to government and to the group]. Hard to get people to come to meetings [due to the distances involved]. Need more young people to be involved. Lack of time now [to get involved].	Public awareness of work – more people have joined. Big impact on government, official recognition from them. Provided a role model for governments. Cross-cultural issues raised – re Aboriginal community. Logo competition involved children. The group is together and talking.

Involvement and awareness

Involvement means many different things to different people. Robbo from Water Watchers said: "involving kids was very successful, proving the adage 'Tell me and I'll forget, show me and I'll remember, involve me and I'll understand'". Certainly for this group, school children, land holders and residents were "involved" in water quality monitoring at the sampling stage, but not everybody took the process through to performing calculations, drawing conclusions and developing action plans.

Using a different interpretation of involvement, Billie from the Downside Landcare group felt involved because: "you know what's going on in the area, and you get to know the neighbours better through being involved with the group". Lance and Daphne also spoke about more involvement with their neighbours in this positive way. Personal satisfaction and enjoyment gained from stewardship group involvement was very important for Downside Landcare group members:

> I enjoy seeing the trees growing that we've put out. It's a feeling of self-satisfaction and achievement, and I've met some really nice people. (Lena)
>
> A lot of satisfaction, togetherness, a feeling of camaraderie, some hard work, some worries, no regrets and I've learned a lot for sure. (Kevin)

In all three groups, members were most proud of the involvement of a wide range of different types of people in stewardship projects, whether they are home makers or violin teachers, graziers or bush stockmen. This diversity of involvement was particularly apparent in the Water Watchers and Mitchell River groups and led to increased involvement and awareness.

Greater awareness of an environmental issue or problem coupled with increased local involvement are positive outcomes of stewardship groups whatever their focus. The range of activities designed to improve awareness and involvement was impressive in all three stewardship group cases. Water Watchers hosted a booth at the local show, the Mitchell River group erected a display at the open day of a large gold mine in the catchment. The Downside Landcare group hosted a dinner for a group of visiting New Zealand farmers. To share costs and responsibilities, local group members cooked at home and delivered the hot meal within minutes to the waiting guests at the Tennis Club hall. Hosting visitors was their preferred public involvement strategy, but the group were also very experienced in conducting local field trips and regional bus trips.

The Mitchell River group routinely held half-day field trips before or after their scheduled meeting. These trips were often combined with a social event. Due to the vast distances in the watershed, most group members travelled to the meeting site the night before. On one such occasion at the Kowanyama Aboriginal Community, the group hosted a "kubmurri" (an underground oven similar to a New Zealand hangi) which in itself was a cultural exchange and proved to be very successful in terms of raising awareness and encouraging involvement. Elders from the Kunjen, Kokoberra and Kokomnjen groups spoke of historic battles and shooting incidents between white settlers and Aborigines. The elders welcomed Mitchell River group members to their country and underscored the importance of cooperation and understanding in cultural aspects of land management.

Greater awareness was one of the most frequently cited examples of personal achievement in environmental stewardship.

> [We've] built an awareness of ecological [matters at] meetings. Now people know about salinity. The first meeting 3 people put up their hand to own up to salinity. At a later meeting about 10 people put up their hand. (Mike, Downside Landcare group member)

> [There is] much greater awareness and knowledge of phosphate content in streams and drains with which to make more intelligent decisions in the future – [it has been] a tremendous insight into problems in the catchment. (Robbo, Water Watchers group member)

"More awareness" is a tangible product or result of stewardship groups' efforts. Mitchell River group members spoke of the importance of a brochure as an identifying feature of the group. Blane and Brian both mentioned articles in the *Cairns Post*, talks given at Rotary meetings, information stands put up at horse races and rodeos, field trips, the role of a new logo and the necessity for signs to be erected.

The Downside Landcare group created awareness through newsletters and through their efforts at television and newspaper coverage. Despite all these efforts there is still a marked trend for stewardship groups to down-play their role in raising awareness, in favour of an accent on specific projects and *doing something*. It is worth emphasising that there is always a close association between heightened public awareness and individuals taking action:

> [I'm] more aware of wet spots and problem areas – you can sort of tell the spots to look for. [I have] more awareness of other people's problems from bus trips etc and [you're] more conscious of preserving your own soil. I have planted trees due to [my] awareness and have sown phalaris on the hill. (John, Downside Landcare group member)

Stewardship action

"Runs on the board" in terms of getting results and doing something concrete is a common concern of stewardship groups. The three groups illustrate different phases of group development and different approaches to stewardship (see Table 4.2). The Mitchell River group was more holistic in their approach. It actively considered the more abstract notions of intergenerational equity, sustainable development and the need for a community-based cooperative approach to stewardship. Beyond their goal of "looking after" the catchment, group members were particularly concerned with group establishment and process as a first step toward environmental stewardship. Whilst Gareth and Jim perceived the lack of specific action "on the ground" as a weakness, others considered that the group had far more potential to effectively design and implement specific projects having paid due consideration to matters of process and "theory" first:

> I'd be satisfied if nothing ever happens to the Mitchell River – just for the group to keep an eye on it. The message for the Queensland Government is that it is good that the group was established before problems occur. (Barney, Mitchell River group member)

Members of the other two groups were quicker to point out their specific stewardship achievements. Their primary activities can be classified as monitoring and vegetation projects.

Toward Water Watchers members' goals of measuring phosphorus loads in water quality, the group undertook three main activities; collecting water samples for analysis, recording flow rates and making a detailed profile of the drain or stream where the sample was taken. Based on this thorough monitoring process, members compiled a research plan of where to monitor the

Table 4.2 Stewardship action

Bad points	**Good points**
Water Watchers	
[Everyone] sampling on the same day was difficult. Picking a day suitable for all schools was difficult. Telephone tree system broke down. Teaching everyone how to do cross-sections of drain profiles was difficult. Samples were not labelled correctly. There's no information about how to catch the [phosphorus] slug. Some drains were dry and others were not. Checking water velocity was difficult when there were weeds clogging up the drains.	The sense of achievement, beyond our expectations. Enthusiasm about learning new things. There was a big range and some very interesting results. A sense of involvement in taking action.
Downside Landcare	
The time taken before you see results – most of us may not see results. People don't feel they have enough time. [There's] guilt that we haven't come up with anything that will help everyone.	Demonstration sites prove what you have done. Good first meeting called – everybody came. We've strengthened the social structure of the district through tree planting and building the greenhouse. We've had minimum-till demonstrations. We've proved things with lime trials and demonstrations. We've lobbied for rebates on freight charges for lime.
Mitchell River	
Problem of distance. No regular face-face-contact. Hit and run meetings.	

following year so as to maximise their chances of identifying point source emissions:

> [We've] already found out and clarified point sources and general phosphate hot spots. It has pointed to subcatchments which need further or greater monitoring. [We've] found that market gardeners may not be contributing [many phosphates] and that wetlands were stripping phosphorus from the water. (Robbo, Water Watchers group member)

Unlike the Mitchell River or Water Watchers groups, the Downside Landcare group specialised in vegetation projects. They established a nursery for growing and cultivating tree seedlings to plant on salt-affected properties and thereby minimised the cost of buying trees. They also had regular "pricking-out" days or working bees to subdivide and transplant growing seedlings. Members involved a local primary school in collecting seed from endemic species and had regular tree-planting, fencing and weeding sessions where the whole group had the opportunity to participate. Other group activities included implementing a Green River Banks project in conjunction with Greening Australia and the establishment of lime-strip trials, pasture demonstration sites and cultivation practice demonstrations.

Education and communication

As with increased involvement, education was similarly applauded as a great benefit of stewardship group action, but criticised as being a "soft" rather than a "hard" product. Like involvement, increased or improved opportunities for learning could reflect both a personal achievement, or a whole-of-community gain (see Table 4.3). Unlike public involvement however, members' descriptions of their own environmental education could be easily identified as a product arising from a specific stewardship group.

All three groups involved educational institutions to some degree in environmental education projects. The Water Watchers group actively involved children from five local primary schools monitoring water quality in nearby creeks and drains. Cathy visited these schools and spoke to children about the autumn "slug" of phosphates flowing through the waterways at the end of summer. This concept captured the children's imaginations and served to increase their understanding and awareness of ecological relationships, specifically those pertaining to nutrient enrichment issues.

Table 4.3 Education and Communication

Bad points	Good points
Water Watchers	
It is too soon to judge whether the results can be used. Maybe people will not take any notice of a report. Who will read it? There are no funds [to act on the results of the report].	We had a lot of support from the Community Catchment Centre. There was a report produced which is professional, credible and readable.
Downside Landcare	
Very complex problems. The [high] cost of knowing, eg with fencing. Some people know but don't do anything. We haven't come up with something from the demonstration sites.	Good learning. Kids will be running the country in the future. Kids will be influencing adults. With demonstrations we are learning from each other.
Mitchell River	
No comprehensive education project. Some people don't get newsletters or don't read written information.	Improvement in discussion. Getting stakeholders together. Appreciation of other stakeholders' problems. Talking *with* not talking *at* people.

Since the Downside Landcare group's inception, it has forged links with the Wagga Wagga Technical College of Tertiary and Further Education, the Marrar Primary School and Charles Sturt University. The latter two groups undertook fencing and tree-planting projects as part of their course work on several landcare group members' properties including Mike's. This not only deepened students' understanding about landcare, it also contributed to awareness of salinity problems throughout the broader community. The Marrar Primary School invited the Downside Landcare group to their school to assist with tree planting. Later, the school invited the group to breakfast on Melbourne Cup day in recognition of the group's assistance.

The Mitchell River group also started environmental education projects with schools in their area. However, they were less advanced in this activity due to the huge distances involved, and their early stage of development as a group. A logo design competition in local schools may have served to increase community awareness of the group itself, but probably did not go far enough in environmental education terms.

However, educational outcomes are not restricted to school children's involvement. Many stewardship group members refer to critical and interactive educational outcomes arising from adults' experiences of field days and workshops. Much of this increased learning assumes a quite personal tone. Billie from Downside said: "I've learnt to apply knowledge to my own problems on the farm – then you know what's going on in the area". The primary vehicle for environmental education for adult members of stewardship groups is that provided by a social learning context, getting together over BBQs, cups of tea and the chance to have a chat.

Richard, the facilitator for Water Watchers would spend time with group members one to one, as well as in workshops explaining how to sample and monitor water quality. Drain profiling, water sampling and calculating phosphorus loads and stream flow rates are not particularly easy tasks for those untrained in science. The Community Catchment Centre (where Richard was based) proved very helpful to Water Watchers in providing the technical information and practical skills necessary for the group to complete these tasks. The Centre's unwritten philosophy was "science for everyone".

The Downside Landcare and Mitchell River groups relied heavily on government assistance with their workshops and extension services. This affiliation with government agricultural and land management agencies was crucial to the way both groups worked. One of the most successful of these associations for the Downside Landcare group was with officers from the New South Wales Department of Conservation and Land Management who conducted workshops on whole farm planning. Similarly the Mitchell River group liaised with the Queensland Department of Primary Industries to undertake a workshop on dryland salinity at the top of the Mitchell Catchment. Without these workshops, stewardship group members would have had far fewer opportunities for information exchange and informal learning.

The extent to which stewardship groups define themselves as providers of information for the broader community will vary enormously. Some groups take on information provision as a primary role in raising community awareness. Others will publicise their activities and regard information dissemination as an indirect outcome of what they would have done anyway. Typical strategies for disseminating information include the use of public demonstration sites and publication of formal reports.

The establishment of demonstration sites and on-farm trials is one of those activities in Australia most associated with landcare groups. One of the Downside Landcare group's first projects was to plan a demonstration site on a saline recharge area. The ground was mounded, then planted with a variety of local tree species. A piezometer (a narrow cylinder inserted down a hole in the ground to measure the height of the water table) and interpretive signs were used to demonstrate groundwater height so that passers by would have a clearer understanding of dryland salinity. Subsequently, demonstration sites were set aside on Kevin's and Mike's properties showing different approaches to combating salinity and improving productivity. Cropping trials, minimum tillage trials and lime trials have all been used by the Downside Landcare group. This information dissemination process seems to have benefited a wide range of landholders in the community apart from those on whose property the trials and demonstrations have occurred.

The Mitchell River group have also undertaken demonstration and interpretive exercises. In Chillagoe, the group worked on a project designed to halt further degradation of an eroded gully. They prepared a demonstration site and put up a sign near the main road entering the town to interpret the various land degradation control measures. The Mitchell River group also erected interpretive booths at local events such as the Red Dome Gold Mine's Open Day, the Chillagoe Races and the Kowanyama Rodeo. However, as Brian and Blane pointed out, with limited human resources stretched over a large geographic area, stewardship groups in remote places should not rely on demonstration activities as their primary method for disseminating information. A more reliable strategy is to publish reports of their work - either in writing, or using local electronic media.

The Mitchell River group produced a brochure which served to publicise the group's existence and raise awareness. They also produced a series of Short Reports - two to three pages long, which identify issues of concern to the group. These publications served to promote the projects of the group and to identify the group as a collective entity in the wider community. A similar series of written reports was released by Water Watchers.

The end of year Monitoring Report compiled by Richard for members of Water Watchers, details the importance of collecting information to form a base of local data on which to build action plans. The document reports that data collected by Water Watchers is being used in at least two ways; in the Shire's Rural Strategy and by local landholders interested in farm planning. Water Watcher's report influenced the shire to such a degree that they rethought their rural strategy in light of the results. This level of influence by stewardship groups, on their local government and wider communities is critical "evidence" of the sort of change these groups can facilitate. However, not all of the results of such group work can so easily be quantified.

Summary

New managerialists have long attempted to come up with a "recipe" for successful group practice. Issues of leadership, conflict resolution, meeting procedure, and so on, have been scrutinised carefully by consultants of organisational behaviour to determine the "right mix of ingredients". Most of these people then turn to examine group outcomes in detail in order to illustrate their particular theorem or hypothesis about group practice. But surely success cannot be measured by such a mechanistic approach. In all three cases, group members spoke of the importance of attending to *process* as well as to products and mentioned a number of unexpected, or less tangible outcomes. Take for example, the community spirit or pride in the district which many people associate with landcare. Group members saw Downside as not just a nice place to live, but as a cohesive community built upon a series of informal ties made more apparent with the advent of landcare. Although there were other local networks before the landcare group started, such as attending church or playing tennis, for those who were not linked into either of these, landcare proved invaluable as a way of identifying with other community members.

Both the Water Watchers and Mitchell River groups also reported unexpected or unintended outcomes arising from their work. Government departments became aware of community initiated projects and as a consequence money has been directed to assist new groups form, or to generate informative brochures. Robbo said that as a result of their water quality monitoring activity:

> It has shown up terrible gaps in the Environmental Protection Authority's and the Water Authority of Western Australia's monitoring programs. Many of their statements were based on assumptions, not facts.

Another important but unintended outcome for these stewardship groups is the increasing interest in the environment from a wider cross-section of the community. Attention to the stewardship movement in Australia has also been increased through government networks of landcare, water watch and catchment management coordinators in every State or Territory. However, while interest in stewardship is not in itself a bad thing, unchecked academic or government attention to rural communities can lead to burn out in some groups and perhaps contribute to an artificially high level of self-awareness in "target" communities.

Despite new managerialism, it is appropriate to consider the phases of group development in relation to the three stewardship groups. Theories of group dynamics illustrate that there is a series of developmental phases that groups go through between formation and maturity. These phases have been characterised by management experts as: forming, storming, norming, performing and mourning – indicating the types of group dynamics acted out in each phase.[7] The language and philosophy of community development practitioners on the other hand is more attuned to voluntary stewardship groups. In a study which specifically related to landcare groups (but could also be important for stewardship more generally), it was suggested that group phases follow a "participative action model": [8]

1. **Emergence:** individuals notice problems, issues get discussed, ideas are generated on how to handle the difficulties, meetings and workshops are held.
2. **Establishment:** the group is founded as an entity unto itself, leaders emerge and members adapt to the style of leadership demonstrated, organisational structures and project plans are discussed.
3. **Action:** members work cooperatively, and complementary skills and knowledge are recognised, resources are gathered, projects are implemented, mature relationships

develop and the group may evolve through different leadership and structural arrangements.

4. **Expansion or extinction:** groups either undertake new projects and functions, or become redundant when their current projects are completed. Either way, the group undergoes a transformation.

Water Watchers undertook several cycles of monitoring water quality in the action phase, but then declined in membership, resources and energy before disbanding. The Downside Landcare group evolved through all four phases and has subsequently expanded its organisational structure and project development. The Mitchell River group spent a long time in the process of moving between establishment and action phases and have subsequently released their documented catchment management strategy. All three groups successfully negotiated issues of group process. The importance of group practice to group function is illustrated in Table 4.4. Items listed as being of major importance to group function can be interpreted as having a significant detrimental or beneficial effect to the overall "success" of any stewardship group.

Table 4.4 Importance of group practice to group function

Processes	WW	DL	MR
Individual members			
Motivation to join group	■	■	⬛
Skills	⬛	▪	▪
Attributes	▪	⬛	■
Factors restricting involvement	⬛	■	⬛
Group structure			
Composition of membership	▪	▪	⬛
Number of members	▪	■	▪
Leadership	⬛	⬛	■
Conflict	▪	■	⬛
Decision-making	⬛	■	▪
Representativeness	■	⬛	⬛
Education and training	⬛	⬛	■
Goals	⬛	■	■
Projects	⬛	⬛	⬛

Key: ▪ Minor importance ■ Some importance ⬛ Major importance

Endnotes

1 Good examples of such literature include Chamala and Mortiss' 1990 book and more recently Chamala and Keith's 1995 book. For a more academic and critical examination of group issues within a landcare context see Lockie and Vanclay's *Critical Landcare* published in 1997.

2 Garkovich 1989: 209. Indeed, Curtis 2000: 59 goes further to suggest that landcare groups can learn a lot from the literature on voluntarism. not only about leadership but also, "about how to attract, retain and coordinate the efforts of Landcare members; secure the support of other community groups; structure agencies to support Landcare, manage the inevitable conflicts between Landcare members and agency staff; and undertake evaluation of Landcare".

3 Dempsey 1990.

4 See Gray 1991, 1992; Crombie 1995: 70; Gray, Dunn and Phillips 1997 for a related discussion about power and leadership in rural communities.

5 Gray 1992, makes this observation based on earlier research conducted by Buller and Hoggart 1986, and Newby et al 1978.

6 Quaid Projects Management Brochure 1993.

7 Watson, Vallee and Mulford 1981; Chamala and Keith 1995.

8 Chamala and Mortiss 1990; Chamala and Keith 1995.

5 Red tape or green?

Government involvement in environmental stewardship

This chapter examines government support for community-based stewardship and explores questions such as; who is involved; how is that support manifested; and what roles do governments play in fostering stewardship? While it is generally accepted that governments have traditionally supported stewardship groups, both financially and in-kind, increasingly these groups are supporting and enhancing the ability of government agencies working in this area. However in Australia, not all government agencies are comfortable with this trend, resulting in confusion about the nature and degree of government support for local environmental stewardship. According to Phil, "it depends" on the goals and values of government with regard to the environment and on stewardship groups' requirements for independence or guidance:

> Governments have to sit down and ascertain "if this group is worthwhile" - and then "how worthwhile is it?" - and then, depending on the degree of benefit to the community, give appropriate support ... It [also] depends on the group and whether they want government officers to be advisory or right in with it, boots and all. (Phil, Mitchell River government officer)

Some government officers are uncertain about their role in relation to community stewardship activities - questioning whether these are top-down or bottom-up initiatives. For those who are clear, a variety of roles are available to governments including: resourcing, facilitating, informing and advising, consulting, regulating and evaluating. These roles are all discussed below. First it is important to consider the policy contexts in which government support of stewardship groups has come about generally, then with specific reference to Water Watchers, Downside Landcare and the Mitchell River groups.

Policy contexts

It is useful to consider national and international policy[1] trends on community participation and involvement in environmental management. First, the global drive for sustainable development, discussed earlier in Chapter 2, undoubtedly affects many governments' policies and encourages more community involvement in decisions about local environmental management. In Australia, both State and national governments have been major players in developing the National Strategy for Ecologically Sustainable Development (ESD) and that process involved industry and non-government organisations. There has also been considerable emphasis placed on the 1992 Intergovernmental Agreement on the Environment which establishes principles for coordinating environmental policies among all three tiers of government: Federal, State and Local. Some believe that ecologically sustainable development in Australia would be implemented more often if there were more localised legislative frameworks for environmental management, akin to those existing in New Zealand under the *Resource Management Act* 1991. However, the differential experience of Resource Management Councils in New Zealand suggests that strong local economic bases are also vital to councils in terms of their capacity to enact ecologically sustainable development. Given that both economic and legislative frameworks for ESD are closely tied to government structure, the three tiered system of governance in Australia with its attendant complications means that ESD in this country is still better in theory than in practice.

Principles of integration, participation, partnership and coordination are also cited in national and international policy contexts as evidence of government support for community involvement in environmental management. In Australia they have been voiced in the National Conservation Strategy for Australia, in the National Strategy for Ecologically Sustainable Development, the Inter-governmental Agreement on the Environment[2] and more recently in a plethora of promises to "the environment" arising from the sale of the telecommunications corporation, Telstra. Despite these national and international agency mandates for government support of local environmental stewardship, what happens at the grass-roots occurs more frequently because of the onset of a complex environmental problem than as a result of an abstract process of policy establishment since "it is usually the environment that takes the initiative, often in the form of an extreme signal to the community, such as a bloom of blue-green algae".[3]

In this era of reactivity, nations the world over take action based on a sense of environmental urgency – whether it is perceived by many or few – more often than they act in response to a considered appraisal of what is important to our collective environmental future. Perhaps that is because we have run out of time to sit and reflect. Catastrophic events are occurring more frequently and with greater intensity.

Unfortunately, for many years it was not until after a problem had appeared or a management plan had been drawn up, that decisions to involve key stakeholders were made. In the last third of the 20th century, governments had become increasingly sensitive to the public apportion of blame for environmental damage and simultaneously had strengthened their economic interest in reducing environmental impacts. As a direct consequence, they established a whole array of policy goals inviting public involvement including environmental impact assessments (EIA) and social impact assessments (SIA), local round tables and catchment management trusts, mediation and negotiation forums, green product endorsements and a focus on environmental education programs. Yet, because of time restrictions coupled with a tendency toward ahistoricism (not learning from the mistakes of our forbears), these goals are often forgotten.[4] Or perhaps they've been conveniently misplaced in the continuing quest for economic reform and rationalism. Public involvement is, after all, an expensive addition to traditional approaches.

Given this sense of urgency, it is apparently difficult for governments to act before a problem presents itself. Some of the factors which make it difficult to develop pro-active environmental policies include: the recognition that socio-political time-frames are completely out of sync with ecological time-frames, pervasive scientific uncertainty; lack of data; lack of monitoring and evaluation systems and lack of feedback loops. Internationally pro-active environmental policies are important attempts to involve more people and make better collective decisions, but even they are "vague, lacking measurable goals, overwhelmingly non-binding, and [are] not underpinned by substantial institutional or statutory provision".[5]

Government programs in support of stewardship activity are rarely developed in a simple linear or chronological fashion and almost never pro-actively. It seems they are developed more as a result of a community push for an increased say in the provision of social goods and environmental protection than as a result of a concerted government pull toward future sustainable development. Programs, projects and policies develop iteratively and are often borrowed from other regions, other States or other nations as was the case for the Water Watchers group (described in more detail below).

Whether community-based environmental stewardship is an abrogation of government responsibility, or a means for local people in degraded places to get involved and assist in the process of environmental restoration is a matter of interpretation of specific circumstances along a continuum. The essential tension lies in whether one views environmental restoration as a feature of the classic paradigm of top-down, State-initiated science complete with technology transfer processes back to local people and places; or as a feature of more recent neo-populist thinking which emphasises bottom-up participation, collective learning, empowerment and socio-political "solutions".[6]

Somewhere between top-down and bottom-up approaches to environmental management is a place called Regional Australia, also known as the middle ground or the great unknown and it is fast gaining political credibility as a cross between a black hole and a hot potato. I don't want to dwell upon the pros and cons of regionalism, that has been done before. What I do want to explore is what regionalism means in an applied environmental context when looking at the distribution of funds from the Natural Heritage Trust (NHT) – the Australian government's (rapidly dwindling) conservation program for land, vegetation, rivers, coasts, marine life and biodiversity. That is, I want to investigate regionalism embodied by regional assessment panels (RAPs) who make recommendations to relevant Ministers regarding NHT funds. As set out in the Commonwealth-State/Territory agreements on the operation of regional assessment panels:

> Regional and State assessment panels will be chaired by a community representative and have a majority of community membership. Assessment panels will encompass a broad membership with skills or experience covering environmental protection, (including biodiversity conservation), sustainable agriculture, natural resources management and, where appropriate, land, water, marine, vegetation, conservation, farming, indigenous land management, and State and local government.[7]

Regional assessment panels and their processes differ between States and Territories. Some State processes are more government oriented, others have a more community-development goal. In other words, the middle ground is not always in the middle. Most RAPs are somewhat closer to the top-down than they are reflecting bottom-up issues and priorities. How each RAP is selected and supported, and how each interprets its function, illustrates this difference.

The number of RAPs in each State or Territory is not consistent across Australia. Some are bounded by social, historical or economic divisions, others by catchment boundaries. Victoria, for example, has 10 different regions (and therefore 10 RAPs over its 227,600 square kilometres) and has developed priorities reflecting critical issues within each region. Western Australia, on the other hand, has only six RAPs across 2,525,500 square kilometres, and they are based primarily around land-use and land capability distinctions – not strict catchment boundaries.

There are also different numbers of representatives on each RAP, selected using different criteria according to very different selection processes. Some States placed a public newspaper advertisement, others called for nominations or expressions of interest in the positions, and others asked existing organisations to nominate representatives. Most States and Territories have a secondary process whereby those members nominated are "signed off" by the appropriate Minister or Ministers. There is also an unstated consensus between States that the selection process should be based upon individuals' skills and experience rather than their organisational allegiance. There appears to have been a concerted effort on behalf of most States to widen their RAP membership to include people skilled and versed in issues of biodiversity, native vegetation and indigenous affairs. However, in one State, little real effort has gone into reshaping these panels in terms of their previous emphasis under the National Landcare Program (NLP) on agriculture and productivity goals.

When it comes to the actual process whereby RAPs make distinctions about which projects should be approved and which should not, there are even more apparent differences. Almost all States have a technical assessment procedure which runs in parallel to the regional assessment process. Mostly this entails another committee of people who meet to assess a project's technical feasibility and projected outcomes against economic outlay. In at least one State there is obvious tension, with technical assessors making disparaging comments about the willingness and abilities of the RAP membership to read and understand the technical comments provided.

There are also numerous differences amongst States when it comes to the scoring or ranking system each employs to assess natural heritage projects. Some use a one page document coupled with a numerical system aimed to rank all projects in their region. Some suggest guidelines, but leave the process open ended for RAPs to make their own more subjective judgements. Some target projects to specific "buckets" of funds thereby effectively negating the NHT's stated goals of integration and collaboration.

There is a need for closer examination and refinement of the RAP process in each State in order for the NHT to achieve its goals. I am far from suggesting that the responsibility for allocating funds for regional priorities from Federal coffers should lie outside a region. Obviously this sort of middle-ground approach is an improvement on past practice and better reflects local needs and issues in the management of natural heritage. Despite the inadequacies and inequities inherent in the regional assessment process, it is also important to remember the honest bureaucrat's line that the government of the day is merely acting on behalf of the wider Australian community to whom they have an obligation to make sure that money is invested appropriately in each region. In other words big brother is occasionally more brotherly than big.

To make the process more transparent, a joint ministerial statement recently announced five key Commonwealth criteria for RAPs to use in assessing potential projects under the *Natural Heritage of Australia Act* 1997:

1. Projects should contribute to long-term on-ground improvements in environmental resources;
2. they are good value for money;
3. they are both technically feasible and appropriate;
4. the intended project or program will not displace existing government funds spent in the area; and
5. the community contribution is expressed in matching funds signifying strong community support.

There have to be controls and methods for ensuring accountability and evaluation of public expenditure on local projects. The difficulty is to marry top-down and bottom-up approaches to administration and monitoring. On the one hand, "friends of" groups, landcare groups and other local committees have long been calling for more money to be spent directly on environmental restoration projects. On the other hand, local people are seldom in a position to make recommendations about disbursing government funds without having a long-term big picture of resource management in their region. Equity and accountability principles may be difficult to implement via a regional funding approach. At the same time NHT funding delivered through the regional assessment process obviously meets an expressed need in the community for more money spent on the ground.

A preliminary investigation of the 1998-99 approved projects for expenditure of NHT funds illustrates how this money was spent "on the ground". There was a grand total of $282,623,904 approved for expenditure in 4,045 projects. The greatest number of projects of any one program were landcare related. The Landcare program also boasted the biggest expenditure accounting for $87,907,789 or 31.1% of the grand total. Across the different States and Territories, NSW attracted the most funding ($69,542,208) or 24.6% of the grand total. This correlates with that State accommodating the largest population, 6,038,696 people or 33.75% of Australia's total population at the 1996 census. Incidentally, NSW occupies 10.4% of Australia's total area, but accounts for a significant proportion of severely eroded land in the Murray Darling Basin.

Approved funding was allocated to 6,569 different organisations.[8] Of these more than half (57.24%) were either Aboriginal organisations or community groups. The rest were comprised of State or Commonwealth agencies, local government, industry groups, private companies, regional organisations, schools and tertiary institutions. Community groups (including Aboriginal organisations) were granted $56,613,852 or 25.3% of the total NHT funding available in the 1998-99 financial year.

Across all projects and organisations, the average amount granted to any one project was $69, 870, but over three quarters of the projects (77%) received less money than that. It is notable that there were 28 projects funded in excess of $1,000,000 and of these, 15 were Local, State or Commonwealth government agencies. 18.6% of the total NHT funds available went to the top 28 projects. For an indication of the 15 projects which received the most and 15 which received the least, see Tables 5.1 and 5.2.

Table 5.1: 15 Least Funded NHT Projects for 1998-1999

Project Title	Organisation	Funds
Dust Suppression Tree Plantings	Kurraya Community NT	$ 280
Kiana Bay Access Management	Lower Eyre Coastcare Assn SA	$ 300
Bogantungan Land Management Record	Bogantungan Reserve Committee Inc. QLD	$ 300
Midlands Gully Stabilisation and Shelter Corridor	Hovells Creek Landcare Group NSW	$ 335
Barbarea australis Recovery Plan Implementation	Department of Primary Industries, Water and Environment TAS	$ 370
Fencing Native Remnant Vegetation "Stony Rise'	National Trust "Mooramong" VIC	$ 400
Manilla Yarran Stands Biodiversity Recovery Program	Manilla Landcare Group Inc. NSW	$ 400
Regeneration Trials of Alpine Vegetation in Central Highlands	Upper Derwent Valley Landcare Group TAS	$ 406
Farm Management of Native Blue Grass Pastures for Dryland Dairy Farms	Subtropical Dairy Program Inc. QLD	$ 486
Monitoring Beach Debris at Southend	Millicent North Kindergarten SA	$ 500
Dust Suppression Tree Plantings	Noradidgee Aboriginal Corporation NT	$ 550
Ocean Based Litter Survey	Robe Professional Fisherman's Association SA	$ 570
Normera Wind and Water Erosion Control	NL and ML Williams SA	$ 600
Aminbidji Community Revegetation for Environmental Health	Ngaringman Resource Centre NT	$ 624
Protection and Regeneration of Parklands in Marree and Alberrie Creek	Marree Arabunna People's Committee SA	$ 637

Source: Adapted from Environment Australia, Commonwealth of Australia, 2000 (http: //www.nht.gov.au/)

Table 5.2: 15 Most Funded NHT Projects for 1998-1999

Project Title	Organisation	Funds
Shepparton Irrigation Region Surface Drainage	Irrigation Committee of the Goulburn Broken VIC	$1,370,000
WA Component (FarmBis expenditure in WA is paid from WA's Reserve of Commonwealth Funds, originally received through the Rural Adjustment Scheme)	Rural Adjustment and Finance Corporation of WA	$1,370,262
NSW Landcare Facilitators Project	Department of Land and Water Conservation NSW	$1,425,995
Upper South East Integrated Catchment Program	Primary Industries and Resources SA and S/E Water Conservation and Drainage Board SA	$1,500,000
Acquisition of Albinia Downs	Department of Environment and Heritage QLD	$1,700,000
Establishment of Nairana National Park Proposal	Department of Environment and Heritage QLD	$1,800,000
Highland Irrigation Districts Rehabilitation and Restructuring Initiative	Central Irrigation Trust SA	$1,840,000
Northern Mallee Pipeline Project – Stage 4	Wimmera-Mallee Rural Water Authority VIC	$2,000,000
Implementation of Berriquin Land and Water Management Plan	Murray Irrigation Ltd – Berriquin Land and Water Management Plan Community Group NSW	$2,366,800
Nature Based Tourism Package	Department of Primary Industries, Water and Environment TAS	$2,395,000
Hexham Swamp	Hunter Catchment Management Trust NSW	$2,700,000
Northern Mallee Pipeline Project – Stages 5-7	Wimmera-Mallee Rural Water Authority VIC	$2,800,000
Wet Tropics Baseline Management	Department of Environment and Heritage QLD	$3,200,000
Property Management Planning Campaign – Farming for the Future Stage 2	NSW Farming For the Future Board of Management NSW	$4,478,851
Bushcare Support Services 1998-99 to 2000-01	Greening Australia Limited FED	$4,955,709

Source: Adapted from Environment Australia, Commonwealth of Australia, 2000 (http://www.nht.gov.au/)

Table 5.3: Selected milestones of public involvement in Australian environmental policy

1974	*Environmental Protection (Impact of Proposals) Act* (Commonwealth) provided for limited (reactive) community participation.
1975	*Australian Heritage Commission Act* developed the Register of the National Estate reflecting many places of cultural and national heritage nominated by the community.
1979	*Environmental Planning and Assessment Act* (NSW) first major environmental framework for public participation in NSW.
1984	Publication of the *National Conservation Strategy for Australia* establishes the grounds for public participation in both government and private sector decisions.
1985	*State of the Environment in Australia 1986* reported public participation in terms of statistical measurements of attitudes and perceptions to environmental problems.
1986	Landcare endorsed and promoted by the Australian Conservation Foundation and National Farmers' Federation.
1989	Prime Minister Bob Hawke's Statement on the Environment, *Our Country Our Future*, made a financial commitment to cooperative action between community and government environmental management.
1993	*Native Title Act*: arguably the most important milestone of indigenous people's involvement in environmental policy.
1993	*Waterwatch*: a national water quality monitoring program, emphasised environmental education and community awareness.
1994	*Living on the coast*: the Commonwealth coastal policy was developed in consultation with public stakeholders.
1996	A nationwide public consultation program took place as an important part of the Ecologically Sustainable Development strategy which resulted in the release of the *National Strategy for Ecologically Sustainable Development*.
1996	*The National Strategy for the Conservation of Australia's Biological Diversity* and the *Draft Wetlands Policy of the Commonwealth of Australia* both acknowledge and encourage public participation in and ownership of environmental issues.
1997	*National Rangelands Strategy* was released after extensive consultation with State and Territory governments and the public from 1993 onwards.
1997	The *Natural Heritage Trust of Australia Act* is dedicated to "repair and replenish Australia's natural capital infrastructure". Cooperative partnerships between communities, industry and all levels of government is a key goal of NHT, emphasising local solutions to local problems.
1999	*Environmental Protection and Biodiversity Conservation Act* similarly promotes community involvement in assessment and management of Australian environments.
2000	The Council of Australian Governments released *A National Action Plan for Salinity and Water Quality*. Based around catchments or regions, it clearly articulates a role for both individual landholders and local communities in designing action plans for natural resource management.

The changing policy culture of Australian environmental stewardship is clearly illustrated in Table 5.3. It depicts an increase in the number of national policies which acknowledge the role of public involvement in local environmental management, perhaps in response to the more sophisticated approach to environmental issues reflected in recent Australian attitudes and opinions.[9]

Although community involvement in environmental management is not new in Australia, the sheer intensity of effort and large number of stewardship groups does indicate a considerable increase in local activity. And yet, now more than ever before, it is apparent that Australia faces complex, intractable environmental degradation on an unprecedented scale. Rather than use this as an illustration of the powerlessness of stewardship groups in the face of such environmental decline, richer explanations of these continuing land and water degradation problems should be sought in the current socio-political context of governments' hands-off environmental policies. Governments' *lack of support* for stewardship activity merits closer consideration before detailed principles of government *support* for stewardship activity in the three case study groups are investigated.

Common policy failings in environmental management include incremental and cumulative errors of omission, bias and fragmentation, or what critics label the "lurching, myopic, episodic and ad-hoc nature of much Australian environmental and sustainability policy".[10] Admittedly this statement refers to the failings of environmental policy more generally, rather than the policy failings of public involvement or participation in environmental management in particular. However, there is much evidence to support the claim that stewardship activity is taking place in a general climate of inadequate government support dominated by representative processes rather than participative processes and characterised by adversarial relational styles rather than collaborative styles.

At local scales, whole community sectors are frequently left out or misrepresented on catchment management boards, coastcare committees, waterwatch initiatives, RAPs and other government facilitated environmental management committees and boards. Governments, their departments and their officers have a duty to encourage participation by those segments of the population who have not traditionally engaged with civil society, especially people from non-English speaking backgrounds, indigenous and disabled people. And in terms of broader scales of policy-setting, governments should develop more opportunities for key stakeholders to play a bigger role in developing environmental futures. Despite public involvement becoming institutionalised through the regional assessment process, it is not sufficient at a national scale. Nor is it an adequate response when corporate research organisations and government committees continue to invite selected "representatives" of certain environmental lobby groups to participate in decision-making processes which appear pre-determined and not open to alteration. In Australia in 1991, the ESD working groups were a good example of an innovative process of policy development. Although sometimes derided as corporatist, at least it was nationally consultative. In Canada, the National Round Table on the Environment and the Economy provides an imaginative and productive source of ideas from a very broad cross-section of that nation's decision-makers.

In the post World War II era, there has been a demonstrable, but gradual, diminution of government responsibility for rural resource issues in Australia and a simultaneous progression in the degree of community ownership of these issues. To simplify a complex topic, consider the following typology as four stages of policy reaction to environmental degradation:

1.	Governments take major responsibility	minimal community involvement
2.	Less governmental responsibility	more community involvement
3.	Governments devolve responsibility	community takes major responsibility
4.	Governments abrogate responsibility	individuals encouraged to DIY

For example, in 1997 with the release of a Save the Bush Tool Kit[11] individual landholders were encouraged to assess and manage the natural resources on their own properties (stage 4).

Whilst it can be argued that this do-it-yourself approach suits landholders' individual learning styles and engenders ownership of land and water degradation, it can also be perceived as an inexpensive alternative to training facilitators and extension officers in ecological processes. The DIY approach is also occurring in the farm forestry sector, where there are few, if any forestry officers still offering one-to-one silviculture advice. This lack of extension is also symptomatic of the new managerialism prevalent in stages 3 and 4 of the typology whereby "generic management skills are valued above knowledge and skills specific to a particular sector".[12] Stage 4 is especially disturbing since it appears that the current fix-it mentality is not simply a momentary lapse of attention to environmental stewardship but an indication that governments are now actively looking in the other direction.[13]

Whatever political and regulatory responses are made to environmental problems, they cannot be implemented "across the board" without reference to local conditions. The old command and control style of governmental should be, and is being, phased out. Alternative methods of government intervention in local places or local issues are required.

Policy context – water quality monitoring

Water Watchers formed just prior to the evolution of the Western Australian Ribbons of Blue program in which school children monitor water quality as part of their educational curriculum. This was well before the Federal Government's 1993 Waterwatch initiative and the National Water Quality Management Strategy in 1992. There were several factors which pre-empted the group's formation. First, local residents believed that they were being blamed by government officials for nutrient enrichment. Second, the Office of Women's Interests in Western Australia were interested in establishing cooperative programs between local women's interests and environmental issues. Third, various "watch" activities occurring in other parts of Australia had set precedents for involving both adults and school children in monitoring some aspect of the local environment, eg Saltwatch and Frogwatch. These in turn had grown from environmental education and community monitoring initiatives in the United States and from the establishment of the Global Rivers Environmental Education Network (GREEN). Concurrently, the Western Australian government adopted Integrated Catchment Management (ICM) as policy in 1988 after a national workshop on catchment management earlier that year.[14] Subsequently, representatives from the Office for Catchment Management (OCM) assisted Water Watchers to become established. Thus, the group did not grow out of one formal policy context so much as it arose to deal with a problem which a variety of government initiatives were addressing.

Policy context – landcare

Much has been written about how and when landcare started and more especially, who was responsible for its beginnings. Without pinning it down to an actual date, it is more accurate to paint the soil conservation atmosphere in which farmer groups formed, which in turn led to landcare. Soil conservation boards formed as early as 1938 in New South Wales and 1940 in Victoria. These became Soil Conservation Authorities in the 1950s, but it wasn't until 1983 that the National Soil Conservation Program was launched to prevent and control land degradation. In 1985, the *Commonwealth Soil Conservation Act* focused national attention on the plight of the land. Then, in an historic union in 1989, Rick Farley, of the National Farmers' Federation (NFF) and Philip Toyne of the Australian Conservation Foundation (ACF) negotiated a national package of support for Landcare. Out of their collective concern (and based on farmer group experiences in Western Australia and Victoria), landcare was adopted by the Federal Government. In 1990, the Decade of Landcare was announced and $340m allocated to a program which was to "tackle land degradation and develop more sustainable land management practices".[15]

Since 1990, numbers of landcare groups have been increasing. In 1994 there were approximately 2,200 landcare groups, but as we begin the new millennium that number has almost doubled to more than 4,500 according to Landcare Australia Ltd. Concurrently, other parts of the rural community have adopted "the landcare ethic" and, more recently, urban communities are also committing to landcare. The National Landcare Program distributes money to community groups, State government agencies and others through a competitive grants scheme which is now part of the one-stop shop arrangement central to the Natural Heritage Trust. While landcare groups may not constitute a new social movement on par with the peace movement or the women's movement,[16] they are part of a widespread Australian phenomenon concerned with the interdependent relationship between environment and productivity. A focus on present social and environmental issues can offer short-term, practical steps to community action on land degradation issues which in turn offers a route to sustainability. As well evidenced by the Downside Landcare group, cultural and natural landscapes are interlinked. Over time each has helped to shape the other and will continue to do so. Landcare is as much about caring for community as it is about caring for our natural heritage.

In 2000, Farley and Toyne admitted that a decade of landcare was not long enough and that $340 million had not gone far enough. They pointed to the continuing environmental decline in the quality of Australian land and water and losses in soil fertility and biodiversity. Importantly, however they do not believe that landcare has failed.

> In our view, the Landcare movement has created new community norms and attitudes about sustainable use of resources and what it is to be a good farmer. The best landcare performances have been inspirational beyond the positive impacts on the individual properties involved. It is possible to glimpse the land management of the practices of the future by looking at any of the winners of the national and State Landcare awards. The Decade has led to more targeted research, greater availability of information, pooling of resources and significant changes in management practices - above all, a change in the ethic of land ownership towards stewardship.[17]

Policy context – catchment management

In contrast to the landcare movement, much less has been written from a social science perspective about integrated catchment management (ICM). State programs for ICM exist in Western Australia and in Queensland. In New South Wales, a program for total catchment management (TCM) was the first to be legislated in 1989. Victoria has developed catchment management around salinity control sub-regions and working groups. This is illustrated well in the publication *Salt Action: Joint Action: Victoria's strategy for managing land and water salinity*.[18]

Catchment management has its historical roots in water boards of urban areas dating back to the 19th century and in river improvement boards of the 20th century (eg *River and Foreshores Improvement Act* 1948). Water authorities were established to resolve conflicts between different categories of water users. The most notable example of this is the River Murray Commission set up in 1917. However, it has been comparatively recently that integrated water and land management has been advocated, for example with the establishment of the Murray-Darling Basin Commission in 1985.

The common attributes of integrated catchment management include a systems approach, a stakeholder approach, a partnership approach, a balanced approach and an approach "in which attention is directed to key issues and variables, rather than all issues and variables".[19] Intrastate and interstate government agreements have been putting integrated catchment management at the top of the "to do" list for many public and private agencies engaged in water resource planning and management. By involving local people in local issues, the establishment of catchment committees and groups is seen to be one way of achieving a more sustainable Australia. However, like the regional assessment process for distributing NHT funds, the philosophy and processes by which ICM occurs vary tremendously across the States and Territories. The variety of funding programs available under integrated catchment management reflects the lack of a national, coordinated policy. One common method used to instigate integrated catchment management within a local context is for government agencies and professionals to seek out local resources or to "facilitate groups into existence". Arguably, in NSW for example, Total Catchment Management (TCM) policy arose specifically out of a concern to bring different departmental bureaucracies together, not from a particular desire for a local environmental management perspective.

ICM was started in other States by local people. They established catchment groups and committees independent of government, and then invited government representation as technical advisers. Some ICM committees are comprised of representatives of every conceivably relevant government department, industry sector and community faction. Others are smaller, may not call for representatives, and comprise, for example, community irrigators independent from government departments. The permutations and diversity among integrated catchment management models is endless and reflect local concerns. However, integrated catchment management does not really boil down to an institutional arrangement:

> If integrated catchment management is to be successful in Australia, it will have to be based on a bottom-up approach, in which individual landholders, each doing his [or her] own thing and cooperating with his [or her] neighbours in sub-catchment and catchment groups, form the essential foundation of a pyramid structure reaching up through catchment and regional levels to State and Federal Governments. Australians

> have always prided themselves on being do-it-yourselfers. Catchment management, more than any other aspect of water and land resource management, must, in the final analysis, and despite all the rhetoric about cooperation and coordination, be very much a do-it-yourself business.[20]

The problem with the pyramid analogy is that although it offers a broad base of participatory democracy and is grounded by the knowledge and experience of everyday farmers, it reaches up, not only to regional, State and Federal governments, but also to a global network of transnational corporations and capitalist agriculture. When the pyramid is inverted, the weight of most global agricultural production processes effectively crushes local initiative via the application of government mandated catchment management legislation and administration. Ruth Beilin, an Australian academic, was the first to offer such a damning appraisal of catchment management which until recently was heralded unquestionably as a good thing.

Beilin criticises catchment management on a number of levels. Geographically "catchments are originally units of hydrological science. Over the last twenty years however, these have been transferred to administrative land and water management units. Administration and management are by association with hydrology as a science, made to appear scientific".[21]

Democratically speaking, catchment management committees are comprised of appointed officials rather than elected representatives, however they seem to enact the financial disbursement roles of government. Lastly, on an ecological basis, catchment boundaries are also somewhat artificial: "soil types or micro-organisms may not halt at the hydrological line on the map".[22] Whilst catchment management has been a useful model for some stewardship groups, "total" catchment management may not be as well integrated with local cultural practices as originally thought.

Despite this apparently comprehensive policy context in which the Water Watchers, Downside Landcare and Mitchell River groups operate, the question remains: are governments doing enough to support stewardship groups? The rest of this chapter investigates the profile and roles of government involvement with the three stewardship groups.

Profile of government involvement

There is usually a strong relationship between group members and local government officers and between group members and State government agencies situated regionally. Relationships between a local group and Commonwealth government officers and agencies tend to be less common. This is despite the fact that much of the financial support for community groups originates from the Commonwealth, either directly through programs such as Landcare, Bushcare and Rivercare, or indirectly through a cost-sharing arrangement with individual States and Territories. Relationships have been primarily State oriented because land and water resource management functions are primarily State responsibilities under Australia's constitutional arrangements. At different times within the history and operation of the three stewardship groups, members have drawn from the expertise and resources of all three levels of government. Notably however, the compilation of different departments and agencies involved with each stewardship group differed enormously. Consider the individual circumstances of each group.

Water Watchers

- Western Australian Office of Women's Interests (OWI)
- Western Australian Office of Catchment Management (OCM)
- Victorian Salinity Bureau (VSB)[23]
- Serpentine Jarrahdale Shire Council (SJSC)
- Western Australian Department of Agriculture (WADA)
- Western Australian Water Ways Commission (WWC)
- Australian Bureau of Meteorology (BoM)
- Water Authority of Western Australia (WAWA).

When Water Watchers received financial assistance from the Western Australian Office of Women's Interests, it was an unexpected source of funds for an environmental project, but reflected the concerns of local women. The Department of Agriculture also played a part in their establishment with officers based in the district at the Community Catchment Centre. The centre has been described as a red brick house on the main street of Pinjarra with "a distinctly non-departmental air about it, and is home-base for people from other agencies".[24] The centre provided information, skills and expertise, but more than anything else officers employed there provided active listening and facilitation skills. Facilitation included minute-keeping, community liaison and provision of technical and human resources. The group's facilitator Richard, served as an administrator or secretary to Water Watchers while he was working in the district and implemented a number of environmental education programs that have had a lasting impact in the district. However, despite all these positive features of the catchment centre, one of the central concerns for Water Watchers' members was a lack of continuity. After Richard came and left, there were other facilitators who did not stay in the job. Each was employed directly by the WA Department of Agriculture as opposed to those funded by government but employed by local stewardship groups. Their departures may have been due to the geographic proximity of Serpentine-Jarrahdale to Perth (and the concomitant allure of the city and "head office"). Some lived in Perth and commuted. The fact that government officers were not promoted into these positions from the surrounding community meant that there was little, if any, real and ongoing commitment to the Serpentine-Jarrahdale area, or to the Peel Harvey region further to the south.[25] Consequently the group lacked a stable organising committee.

Yet, despite all this, the Community Catchment Centre was a very effective bridge between community and government. What Kevin who led the CCC did best was to integrate the attitudes and perceptions of local people with the diverse forms of technical/ scientific, governmental and political support available. Sifting through the myriad of issues and unravelling the history of complex local problems required a unique approach, one which was capable of motivating a majority of the community and capable of establishing goals and implementing action plans that were sustainable and acceptable to residents. In short, there was a need for a localised approach to integrated catchment management. The role of the Community Catchment Centre was critical to Water Watchers through its capacity to liaise with many different government departments, and channel this information back to the community and vice versa.

Downside Landcare

- New South Wales Department of Conservation and Land Management (CaLM)
- New South Wales Department of Agriculture (DoA)
- New South Wales Department of Water Resources (DWR)
- Greening Australia (GA, a non-government organisation)
- Murray-Darling Basin Commission (MDBC, joint State-Commonwealth)
- Junee Shire Council (JSC)

The Downside Landcare group had traditional ties with the New South Wales Department of Agriculture (DoA) and the Department of Conservation and Land Management (CaLM) as it was then. Officers from the Soil Conservation Service absorbed by CaLM in 1991 took the lead role in encouraging farmers from the Downside Agricultural Bureau to go on the bus trip to north-eastern Victoria to look at land degradation. Mike, Ellen and many other Downside Landcare group members remember this trip as the catalyst for group establishment. CaLM has provided much support to the Downside Landcare group since its inception, in terms of administrative support for the facilitator's position, provision of maps, organisation of farm planning workshops and liaison with individual landholders requesting information. The officers who represented these departments (Samuel, Dean and Harry) lived locally and were well respected members of the Wagga Wagga regional community. However, there were still some adjustments to be made in the minds of these officers in terms of the benefits of landcare specifically and group approaches to sustainable environmental management in general. Some still wore their specialist hat to landcare gatherings and insisted that fixing land degradation was a matter of technical expertise and experimentation. While those at "head office" may have appreciated the deficiencies in the linear model of information transfer from research to extension to diffusion, it was difficult for some field officers to appreciate that a paradigm shift had occurred. However, the Downside Landcare group's facilitator Cathy, was no such technocrat. She was not only a landholder in the Downside area, but had well established networks with farming and teaching in the community. In short, she was someone the landcare group could relate to very well and was subsequently hired by the group with funding from the National Landcare Program. Although she has subsequently secured other positions, she has never cut her ties with the district and, whilst acting as facilitator for the group, did not take responsibility away from the group. Instead, she encouraged members to seek out other departments through which the group diversified and strengthened their contacts with government.

Mitchell River

- Queensland Department of Lands (DoL)
- Queensland Department of Primary Industries (QDPI)
  - Fisheries Branch
  - Land Conservation Branch
  - Water Resources Branch
- Queensland Department of Environment and Heritage (DEH)
- Queensland Department of Business, Industry and Regional Development (DBIRD)
- Queensland Department of Minerals and Energy (QDME)

Government representation on the Mitchell River group was always strong. This was largely due to an initial conference hosted by the Kowanyama Aboriginal Land and Natural Resource Management Office (KALNRMO),[26] which many government representatives attended. The Queensland Department of Primary Industries was a major player in supporting the Mitchell River group via a facilitator employed within the Department (Blane), but also because it is this agency which has primary responsibility for catchment management. Other departments liaised with the Mitchell River group concerning environmental issues such as crocodile deaths, feral pig management, fire regimes and illegal fishing. Mining problems in terms of acid leaching, stream bank erosion, and heavy metal contamination were taken up between the group and the Department of Minerals and Energy. Of particular significance to the group is the control of the noxious weed, rubber vine, and control of the feral pig population. The Department of Lands liaised with the Mitchell River group over local control mechanisms for these exotic species. The Mareeba Shire Council's Planning Department also contributed local knowledge and experience to the Mitchell River group, as did one of their shire councillors.

The Mitchell River group started out with a large contingent of government officers that far outweighed the representation of other community sectors on the group and led to fragmentation of government responsibility. The long list of officers involved were "reclassified" as members of the group and given the same rights and responsibilities as ordinary community members. This was done partially to avoid the dualistic notions of experts and lay members of the group and partially to increase government officers' ownership of and responsibility for the catchment. Unfortunately, not all government officers on the Mitchell River group actively defined themselves as group members, and many did not recognise non-government or community members as the core of this local organisation. They tended not to volunteer their services to the group, and attended meetings only because they had been directed to attend by a superior as a representative of a department. Government allegiances on the whole were therefore not with the stewardship group, although one department representative still purported to be a member of the group when it suited him. Like other government officers feeling threatened by the freedom and flexibility of community group members, it will take time before Mitchell River group's government officers recognise the value of government and community representatives cooperatively managing natural resources. Empowerment of the community is often equated with disempowerment of government agencies by those government employees who feel threatened by the pace and the nature of the changes occurring.

Whose responsibility is it to assist stewardship groups?

If we assume that environmental stewardship is a public good and that stewardship groups are acting on behalf of the broader community, one may also assume that although they take primary responsibility for local environmental stewardship, they should not shoulder that burden alone. Stewardship groups such as the Frogs And Tadpoles Society (FATS) take on the day-to-day ownership of and responsibility for frog protection and habitat enhancement. But "ownership" in this context does not mean exclusive rights and responsibilities, nor does it mean that community based stewardship groups should operate in isolation from the rest of society.

Responsibility for assisting the work of stewardship groups also lies with individual members of society, the public and the electorate. Apart from paying taxes, global citizens have a responsibility to become involved in and keep informed on matters of environmental stewardship. For those who become eco-watchdogs, blowing the whistle on acts of environmental vandalism also implies taking responsibility to help mop-up.

Institutions can also play a role as advocates for and stewards of civic society. The media, our centres of education and our non-government organisations (NGOs) play a big role, despite public perceptions to the contrary:

> Lobby groups like Greening Australia can be doing something visible and positive, not sitting in a paddock squawking at people for doing the wrong thing. (Samuel, Downside Landcare government officer)

Likewise, the private sector generally and primary industry specifically has a social responsibility to stewardship groups to specify product information (for example on agricultural chemicals), to help protect the resource base and to assist financially, not just when they can see that there is "something in it for them". Awards, seeding grants and free samples are all tried and true techniques for assisting groups, not to mention the good publicity that goes with any hand-out.

Having successfully reinvented government to include principles of transparency, accountability and public involvement, PR consultants hired by government departments now claim (loudly and frequently) that everyone is responsible for caring for the land, improving water quality and looking after our river catchments. But there are still questions of who initiates the action, the share of environmental responsibility governments are willing to hand over to external participants, the share of environmental responsibility stewardship groups want and/or are capable of shouldering, and how governments expect local people to manage their stewardship activities given competing demands on their time and resources.

Despite common references to the self-help literature and pleas for help in protecting their dwindling coffers, the bottom line is that governments (both bureaucrats and politicians) have an unmistakable and multi-layered responsibility to stewardship groups. Although it may be old-fashioned to talk in such terms, governments clearly have a moral obligation to help stewardship groups fulfil goals for the common good. They have a social responsibility to include cultural and historical values alongside economic ones in a political process which will always result in unpopularity with at least some segments of the electorate.

Local, State and Federal government departments are obviously responsible for assisting community groups where they have policy objectives that overlap with those of the group. But that role should not be restricted to "helping community groups that are worthwhile and focused on projects that governments think are reasonable" as specified by Dean associated with the Water Watchers group. There should also be careful consideration of projects whose aims and objectives differ from current government policy:

> It depends on the issue, the focus, the intent of the group etc. I think it unreasonable to expect government to support in any way [community group] meetings to plot the overthrow of the government, for example. But for matters that government approves of, then provision of [assistance] would be reasonable. (Shirley, Water Watchers government officer)

Obviously both government and community (stewardship groups, the public, our institutions, corporate headquarters, Uncle Tom Cobbly and all) are responsible for promoting the goals of local environmental management. A partnership approach to stewardship is a logical one.

Assisting stewardship groups: what role do government officers play?

The role and position of any government-funded facilitator for stewardship groups is especially important given the many functions undertaken by this person. At one and the same time, she or he may be perceived as a bureaucrat paid (big money) to further government policy on environmental issues and a local member of the community hired to promote group development. Individual circumstances will dictate how an appointment is made and who directs or monitors an individual's performance - whether that is a community group, a specific government department or a combination of both.

It is the individual government officers who are sought out by group members for support - not whole departments. Frequently there is confusion about which department an officer is representing. Some group members are reticent to admit that governments are offering assistance at all. It is the ease with which group members communicate with and access officers within government agencies which makes the most impact on any group's evaluation of government. The "department" is often seen as an unknown entity, but John Doe is someone who can really help you.

Confusion also exists in government agencies about the exact nature of an officer's role in supporting community-based environmental groups. Two types of assistance are in evidence; that which is called for by stewardship groups (facilitating, consultative) and that which is rendered unsolicited (controlling, directive). The facilitator who plays the former role will be well recognised and respected by local stewardship activists as was Cathy of the Downside Landcare group. This is how Brian and Mike described her:

> Cathy is an inaugural part of the whole group, she puts in a lot of effort. She'd be in the group anyway, (if she wasn't the facilitator) but if she was gone – the group would stumble.
>
> Having a coordinator means that I don't have to do all that tedious government paperwork. Although I've met others who disagree [with employing a coordinator/facilitator], now I can't see how we could do without one.

Members of the Mitchell River and Water Watchers groups also believed that the role played by their facilitator was supportive and facilitative. However, a cautionary note should be sounded for any stewardship group member who becomes aware of a tendency for an over-exuberant facilitator to start taking decision-making power away from the group. It is frequently the independence/dependency question which separates support and facilitation from direction and control:

> We're almost totally dependent on them advising us and we rely on them to regulate others doing [environmental] wrongs. (Phil, Mitchell River group member)
>
> We wouldn't exist without government departments, but I'm really frightened of government departments taking over. (Colleen, Water Watchers group member)
>
> [Government] should stay out of it as much as possible, it should be a liaison role [with] only those involved in a give and take relationship as part of the group, not directing the group. (Robyn, Water Watchers group member)
>
> A government officer can volunteer to do things – up to the point at which he or she takes control away from the group. He or she should not display initiative if it detracts from the group's ability to control its own destiny, or if they are doing too much – getting the group off the hook. The group should do as much as it can for itself – up to the limit. (Tim, Downside Landcare government officer)

There will always be different interpretations of facilitation and different approaches to government involvement in stewardship groups. So it is not surprising that confusion abounds about that role in the minds of some government officers. Glen's remark confirms this: "[you] shouldn't lead them but in initial meetings they might need leadership to get the group started". It may be that government officers' roles at group meetings depend upon whether that officer is a stakeholder in his or her own right:

> It varies, depending on whether the officer has been specifically invited to be a member of that group because of their personal qualities, or is there as a person dobbed in by their department to attend, when they are not interested. (Shirley, Water Watchers government officer)

Some officers think that government involvement in community groups is fine as long as the person involved is invited by the group and does not attend on a regular basis, only when requested by the group from time to time. Others believe that an officer should always be there and, if a specific person is unavailable, they should send a representative in their place. These contradictory perceptions reflect the confusion abounding in government circles as to whether officers are entitled to participate actively in stewardship groups or whether their role is purely passive or supportive:

> It varies with the personal style of the officer and [also] depends on the group's needs. If they're invited as government officers, they should represent their agency and what it stands for and be able to help the community clarify its position and gain the information needed. [But] if they're there as members of the group, there's no specific role, but [you are] not allowed to speak against your department. [It could be] information provision, as long as that is what the group wants. [An officer could] propose suggestions to take the group forward, for example with project management if the group is floundering. They shouldn't give real tight direction to the group, but provide a number of options. They could also play a facilitatory role, not just warm and fuzzy though. [They could] operate cooperatively with other government people there (not confronting [them] in front of the group) – it's a supportive role, behind the scenes thing. And they could take on a clarification role – to help the group see their way forward. (James, Mitchell River government officer)

Government involvement in the Mitchell River group was divided between those who were stakeholders, facilitators or advisers.

Some group members believed that government officers act as advisers to politicians if they sit on community groups. This perhaps illustrates the naiveté of some community members given most government representatives were junior officers, or too far away from the centre of power in the cities. Two government officers I interviewed were quite sympathetic to the community cause. Maxine in WA advocated environmental stewardship to prick the conscience of governments in an attempt to "keep the bastards honest". The other said that government should act to increase stewardship groups' legitimacy.

As with all transitionary, interpretive functions, the role of any government officer is critical to the delicate relations between stewardship groups and government agencies. While community group members emphasise resource provision as the most critical task for a facilitator, government officers tend to underscore facilitation of group dynamics as their most important task. Whatever the perspective, there is still a long way to go in building productive relationships between the two:

> We liken it to two dogs smelling each other, [there are issues of] trust, jealousy, autonomy, and there's room for improvement. Government relations took a lot of energy. There's been reasonable involvement, sometimes we thought it was inadequate and the suspicion and mistrust is still there. But we're gradually getting away from spending all the time building relationships. (Vic, Mitchell River group member)

Despite the distance from full-scale cooperation, there are productive relationships being built between governments and communities. The distinct roles for government involvement in environmental stewardship include:

- resourcing
- facilitating
- advising
- consulting
- regulating and evaluating.

The remainder of this chapter will detail each of these roles.

Resourcing

> Get government involved if you want money, [and if you do that] then a government chappy would be checking how the money is spent – it's just grandstanding by the pollies. (Brian, Downside Landcare group member)

Provision of government resources for local environmental management is vital. Governments make finance available for a wide range of projects (such as the Green River Banks project of the Downside Landcare group or the Mitchell River group's Chillagoe demonstration and interpretation planning project). They also provide equipment (such as water quality monitoring technology in the case of Water Watchers) and wages for group facilitators. However, the contribution that community groups make toward their own longer term security should not be underestimated, nor discounted. A lot of time, money and other resources spent by individual landholders goes unreceipted. That does not mean that their contribution is insignificant. Some stewardship groups (like Downside) have established profitable tree nurseries and there are others who do not rely on government for support at all. Many groups routinely apply to a diversity of funding sources so as not to be dependent upon any one source. It is important to reiterate that all stewardship groups are self-reliant to some degree. The question is: to what extent are government funds a necessary or sufficient condition for local environmental management? Related questions include: How much is enough? What should these resources be used for? Who is accountable for what? These are explored through a detailed discussion about in-kind resources, submissions and training, value added accounting, the rural-urban divide and basic resource use.

In-kind resources

After a landcare meeting held near Wagga Wagga to determine catchment plans and future action for a nearby salt-affected area, a senior government officer said to Cathy: "You could have used the departmental video equipment to record proceedings". Too bad he hadn't suggested it *before* the event. In-kind resources such as the use of office equipment or provision of tools are almost as important to many stewardship groups as the provision of financial assistance. And yet, only the most efficient of government officers have comprehensively investigated the potential for stewardship group use of in-kind resources available from many government departments. Communication equipment, cameras, videocassette recorders, slide projectors and other technology - all are possible resources to an inventive and well-informed community group - but groups need to know what is available to them. Office space, telephone usage, secretarial assistance: all of these in-kind resources could be used more often by community groups. However, user-pays principles are limiting the extent to which government agencies are prepared to permit this type of expenditure. Other limits are the massive funding cuts to government agencies in recent years and the escalation of government regulations on procedures for liability and accountability.

Submissions and timing

Delay in submission processes is another factor that unites stewardship groups in their frustrated quest to draw from the government coffer. The timing of advertisement placement and closing dates should not correspond with busy periods for target stewardship groups, for example, harvesting or sowing crops for those involved in broadacre agriculture. After funding applications have been submitted, the time spent waiting to hear a result and, if approved, waiting for the disbursal of finances is a sore point among many groups. For example, in the Downside Landcare group one bad experience with the timing of financial assistance meant that group members were unsure whether they would be able to carry out their first group tree-planting project. Had funds not been forthcoming in the best planting period in late autumn, the whole project may have been jeopardised. Robyn, Jenny, Louise and Robbo of Water Watchers were similarly thwarted in their attempts to start monitoring water quality as near to the beginning of the wet season as possible in order to catch "the slug" of nutrients washed down creeks and drains with the first winter rains. However, without a definite answer to their submission, group members were not able to instigate monitoring plans as early as they would have liked. Apart from feelings of futility and disenchantment among group members, this resulted in a less comprehensive set of water quality monitoring data.

The departmental perspective on submissions and timing is quite different to that of stewardship groups and depends upon the point within the bureaucratic hierarchy at which this debate is entered. Officers at the community level often contribute to the writing of submissions. At the other end, government officers involved in sifting through the myriad of submissions and applications have to abide by a carefully predetermined set of procedures. If community groups perceive that government bureaucracy is slow, it is often because they do not understand the scale at which these operations take place. Government officers need to explain delays and put in place mechanisms to avoid them if possible.

Value added accounting

Once stewardship groups are successful in securing financial resources and begin implementing projects, they tend to pore over the allocation and disbursal of funds at a level of detail much more precise than that of normal government accounting systems. Also, groups allocate money, implement projects and make adjustments to expenditure along the way with more alacrity and responsiveness than government budgetary systems. It could be argued that group members have a higher stake in ensuring that their money is well spent than the anonymous government employee has in protecting public coffers. Or, it could simply be that governments deal with such vast sums, compared to stewardship groups, that their accounting and allocation processes are simply slower and more unwieldy.

Stewardship groups are careful to make the amounts of money they have go further. One Australian study found that landcare groups' contributions to revegetation in a specific area more than tripled the government's contributions to the same place.[27] Some believe the process of seed funding for stewardship groups can lead to added investment and a multiplier effect in the community. As Delwyn stated, government should provide "seeding money for start-up and a minimal annual financial contribution for administration". But from the community angle trying to match seeding funds with project tasks is a lot like "painful long division". Trying to spread the resources as far and as wide as possible within the community was difficult for Water Watchers, Downside Landcare and the Mitchell River group but led to innovative ways of extending meagre resources.

Rural-urban divides

In Serpentine Jarrahdale where Water Watchers was located, the active movement of financial resources from the more wealthy semi-urban area of the shire to the less adequately resourced southern rural areas meant that more resources were spent in less wealthy areas than would otherwise have been. With the involvement of people from the semi-urban garden suburbs in the north of the shire, many of whom had part-time farming jobs or jobs which were not dependent upon farm income, more time (voluntary labour) was available for water quality monitoring in the south. Through schools involvement and the involvement of more women in the semi-urban areas, the Community Catchment Centre provided a very valuable and largely unrecognised support role which in effect reversed this core-periphery relationship by sending time, money and people to work on environmentally sensitive and fragile rural areas.

The same point about urban-rural divides can be made in reference to the Mitchell River group, with even greater resonance and application. Cape York, despite its high tourism appeal, still contains some one of the more isolated landscapes in Australia. The Mitchell River watershed to the west of Cairns is comparatively cut off the tourist track to the Cape. However, residents at the western end of the watershed in Kowanyama have succeeded in promoting themselves nationally and internationally. They are not unilaterally perceived as an isolated and economically disadvantaged peripheral group dependent upon the Cairns tourist core, but as a self-reliant and relatively autonomous community who are highly successful in the eyes of both State and Federal governments. They also have an identity and profile in the Pacific North West of the United States. The Kowanyama community has helped prove to the rest of the catchment population that by working together, money can be redirected from Brisbane and Canberra to the whole Mitchell River watershed in the hope that entire landscapes will be slowly improved.

Resource use

Patrick of the Mitchell River group expresses a common view that governments could and should financially support community groups by "putting their money where their mouth is". Provision of financial support may well be the most effective means of government assistance to stewardship groups since it buys not only capital items but, more importantly, human resources. In other words, money spent on facilitators' wages means more time invested in local environments. Greg, a Mitchell River government officer, equated government financial support with group legitimacy: "Community groups without government backing do not progress – they make noise and often bite off too large objectives". Henry was vehemently opposed to the expenditure of public funds on projects that he believed were serving few interests in ways which did not always achieve observable results. Compare this attitude with that of Glen, another Mitchell River government representative, who suggested that the best resource use for governments was to give money direct to landholders via groups:

> Local people are the best on-the-spot managers. [They] understand the seasons, understand [local] resources and infrastructure. They should be resourced to provide an environmental management responsibility function from redirected funds from government direct to catchment managers relating to fishing, tourism, [and] environmental functions which are not currently part of economic production.

Phil expressed a third view: that community groups save money for government departments. He espoused the use of voluntary labour from community groups to undertake some of the work of government and allow more flexibility in departmental budgets. Although he agreed that community groups should play an environmental watchdog role, he did not consider it necessary for them to be as well resourced as government.

The potential for governments to harness community groups to undertake government sponsored environmental management as a way of making the departmental dollar go further was an explicit motive of Neville, a Western Australian bureaucrat linked to Water Watchers. He also suggested that community groups were a useful tool of government as a pool of voluntary labour – as long as government was in control of the program objectives. Other officers spoke of assisting community groups on an investment basis – only providing money to those groups likely to show a "return" on funding. These examples indicate that some departmental officers are still missing the point that stewardship groups have a legitimate government mandate to undertake local environmental stewardship ... on their own terms.

The "voluntary labour" enthusiasts may be victims of the widespread trend toward reinventing government in the early 1990s (reflected within new public management) which suggests that governments need to harness energy from the private and non-government sector in order to make their resources go further.[28] One interpretation of this view is that governments are not conscripting power to fulfil their own agendas, they are merely working with and through the legitimate constellation of interests reflected in society.[29] Whilst it may be true that resource constraints facing the state are real and won't go away, non-government organisations are also feeling the pinch. Devolution of governmental responsibilities through coercion and conscription are essentially undemocratic and "inferior to those which serve overall to enhance individual liberty".[30] In many stewardship groups, the balance of power between government and non-government organisations is not particularly even and both parties should be aware of both their own and others' motivations for entering into "partnerships" of this kind – especially in the name of public involvement. It is also interesting to recognise the difference between regulatory and voluntary initiatives toward environmental stewardship. It may be true that government efforts toward environmental regulation require the assistance of the business and industry sectors to promote sustainability. On the other hand, voluntary arrangements between government, non-government organisations and the private sector are not mandatory. Representatives of each must recognise each others' strengths without the exertion of unnecessary power or cooption.

Facilitating

> We're not providing magic bullets or a mental infrastructure – we're providing ways for them to come up with their own solutions. (Richard, Water Watchers government officer)
>
> Facilitation is essentially the art of fostering synergy among others – devolving skills, information and contacts to the people who most need them – land users.[31]

Under the broad heading of facilitating, there is a range of more specific skills and roles necessary for good facilitation. These were all apparent in the styles of the group facilitators in each of the three stewardship groups:

1. Communication skills – both verbal and written skills, but especially active listening skills.
2. Animating skills – involving people, motivating people, arousing a sense of enthusiasm and positive action.
3. Conflict resolution skills – mediating between disputing parties and clarifying areas of difference.
4. Public relations skills – liaising with community groups and the media, setting up displays, workshops and presenting results.
5. Networking skills – linking people with others holding common interests or common problems, sharing information and liaising generally.
6. Planning skills – brainstorming alternatives, envisioning solutions, goal setting, delegating tasks, providing feedback, evaluating actions.
7. Educating skills – involving members of community groups in creating social learning environments.
8. Organising skills – arranging meetings, choosing venues, setting up trips and field days.

In addition to this list, facilitators help groups to identify resources, conduct needs assessments, articulate local problems and priorities and promote the skills development of individual members. The mere presence of Richard who was intent on listening to the Water Watchers group, rather than "advising" them, was seen to be a great asset for the group. Kevin, the coordinator of the Community Catchment Centre put it this way in a conference speech: "The farmers were right, and it was our job to listen to them, and support them".[32] Support in this context is another key variable for effective facilitation. Provision of the physical space and opportunities for discussion are critical elements of facilitation. This alone can allow many decisions to be made by those most affected by a problem or most interested in planning stewardship projects.

Perhaps what unites most facilitators involved in community groups and development practice is their diversity. They come from diverse backgrounds, they undertake diverse tasks and they liaise with people across diverse sectors and territories within communities. There are two central approaches to community development practice. The "task-oriented approach" has arisen largely from a history of locally based economic or employment initiatives reliant on the production of tangible results:

> The attention to outcomes places the community development practitioner in a role of emphasising problem identification, fact-finding, analysis of alternatives, grantsmanship and project implementation. Success is measured in terms of improvement in the institutional and physical environment.[33]

Compare this to the "process approach" that describes the way in which a community operates as a holistic structure, unifying economic with ecological, social and cultural goals. This approach sees group members as resourceful participants in the process and relies less on facilitation as a passive avocation on behalf of a professional and defines it more as an active endeavour of the whole community. Whether task oriented or procedural, the approach taken for each of the three stewardship groups is reflected in members' relationships with their facilitator. It is important to reiterate here that facilitators were somewhere along the continuum from government employee to community member.

Facilitator – group relations

The Water Watchers' facilitator, Richard, considered himself part of the group, not separate from it. Instead of a devolution of skills, information and contacts from the government to the community, at first sight there appeared to be more of a two-way flow between the facilitator and the group. Information from the local land base went to the Catchment Centre and skills and technical expertise were fed back to the group. However, this exchange led to subsequent difficulties for the group when Richard left. Although he described himself as a group member rather than a government employee, it was obvious that he was not a member of the group to the other group members. He did not live locally. The community did not identify with him, and he was paid by the government to belong. The issue of "membership" is an integral one to facilitator-group relations and requires thorough examination for all new facilitators in any group.

Richard developed strong ties with individuals in the group, particularly Louise and Robyn. Furthermore, he was particularly careful to maintain those relationships even at the expense of other duties. However, after he left the group, it became apparent that, even though he had developed and strengthened lines of communication between himself and individual group members, he had not paid attention to the type of relations existing *among members*. Richard had created a wheel of communication with a number of spokes to his hub, rather than a network or web of interconnections. When he left, Water Watchers group members were slow to rebuild group cohesion because they did not have a well developed sense of self-reliance, trust and communication. Water Watchers became dependent upon government facilitation. As Richard acknowledged:

> [I] would have liked to reduce my involvement at the technical end. I would have preferred people to do their own calculations and maybe 30% did do that. But there was a large amount of information to collate – 30 sites and 30 people [through] a central information agency. At the expense of getting information out, overall I would have liked to decrease [my technical involvement] – been realistic about the amount of time it will take – [and encouraged] more local people to be involved.

This inadvertent dependency may also have occurred as a result of Richard's lack of training in sociology, education or planning which are basic to community development practice.[34] Coming from a technical/agricultural background, this facilitation experience was new to him. "With the realisation that the technical focus needed to be balanced by promotional flows ... [I] developed the ideal of local ownership and responsibility at the CCC. [It was] the first time [I'd] used this philosophy". Indeed, had he been able to pay greater attention to his own intuition, the group may have developed a greater sense of self-reliance. As community development theory states:

> The more community issues and problems become translated into technical questions, the more they discourage self-help and broad participation ... Because community problems often become technical problems, community leaders and officials find it easier and more "efficient" to make decisions and take actions without much community input.[35]

With the trend in agricultural extension changing from a one-to-one linear model, to a group approach, it was little wonder that Richard thought a facilitator's role was to: "act as an entrepreneur, between people locally". Importantly however, he did not single-handedly cause the facilitation problems – it was a combination of factors which included water quality monitoring being task-oriented, his appointment by government, and the group's own low skill base and high expectations. Importantly, these same factors contributed to success in the form of a monitoring program that was put in place quickly and provided the necessary data and results. Group dependence upon the facilitator was also apparent in the Downside Landcare group, although not to the same degree.

In the case of Downside, the style of facilitation was less technical and administrative and more grounded in the needs of the community. Cathy was appointed by the group after it had been established a year earlier. Among other tasks, she assisted group members to talk things through, identify local problems, organise committee meetings, involve new members and link up with external sources of advice and information. Cathy operated as just another node in the network, rather than as a hub through which information flowed in and out. This may have been due to her sense of local history and knowledge of the community. Generally speaking, there is an important difference between facilitators who are local and come "bubbling up" through the community, and those who are appointed from outside. Yet whilst it is true that inside appointees have a more thorough grounding in basic geography and community structure, it does not mean that external appointees are always less capable. Consider the appointment of the Mitchell River group's facilitator, Blane. In his case, an outside appointee was the most appropriate, given the potential conflict and entrenched differences in value systems within the group. Again, there are no general recipes or rules to follow. Facilitation style and practice depends entirely upon local conditions and local people.

In the 1980s and early 1990s, facilitation was something done *for* community groups *by* government paid facilitators and was constructed as a passive process. Blane, on the other hand, took a much more active role in developing conflict resolution and mediation techniques between different State departments and local government. A recent graduate appointed from a university town outside the catchment, Blane wrote catchment reports, took minutes, operated a budget, wrote media releases, gave radio interviews, involved the wider community in decision-making, maintained a vehicle and drove long distances across the catchment, which frequently entailed being away from his young family. He interpreted the facilitator's job as "being the legs and the wheels" of the group and he certainly contributed to achieving its aims and objectives:

> I personally have a sense of doing something good – and hopefully [am] fulfilling a role that the community wants filled.

> I'm certainly feeling a personal challenge. I get quite excited when I think of the broader concept of what is going on and having these issues and concerns being addressed at the community level and playing a front line role in the development of things like ESD [ecologically sustainable development]. I have a sense of worth in what I'm doing.

A common characteristic of all three stewardship group facilitators was that they made life easier for the groups. Recognising this central theme of facilitation, the then Western Australian Department of Agriculture established a special in-service training session on facilitation skills for agricultural extension officers. More recently, many such facilitation courses have been offered by both public and private sector training organisations.

Certainly there is a need for facilitation of group dynamics, of information flow, of administrative arrangements as well as for more traditional technical extension. In every facilitator's position, all of these roles are required. However, the extent to which any one approach is taken, whether task-oriented or process-oriented, should depend upon the skill base and context of the community group - rather than upon the facilitator's preferred style. Whether he or she has sufficient personal and governmental resources to be a flexible facilitator versus a navigator or dominator is another distinction between excellent and average facilitation.

Facilitation – navigation or domination

To restate, a facilitator's job entails two key functions - facilitating communication and providing administrative support for community groups. Facilitation is usually based on a philosophy of provision of assistance upon request from the group. In adopting the holistic, procedural model of change, powerful or coercive methods in attaining community development are unacceptable. Instead, the process approach to community development emphasises the acquisition of skills that allow community residents to meet their own needs. However, it seems that government departments are becoming so imbued with the idea of not pushing or directing groups that some officers are no longer able or willing to provide even the most basic practical assistance, for fear of dominating the group. Or perhaps it is because they are already overcommitted. This is especially evident where government officers take on community landcare roles in addition to their normal working hours. There is a fine line between "running the show" and assuming responsibility for decision-making and leadership when it is proffered by stewardship groups. On the basis of the facilitators' experience in the Water Watchers, Downside Landcare and Mitchell River groups, I believe it is often as important for facilitators to put in after-hours work with community groups when they know they have the capacity to assist, as it is for them to sit back and let the group make their own mistakes.

Glen and Phil (government officers in the Mitchell River group) appeared to be somewhat patronising in their attitudes toward community development generally and to non-government members of the group especially. They suggested hiring management consultants to "facilitate people to be more comfortable with groups". Simultaneously decrying community-initiated catchment management on the one hand and attributing the group's positive results to government intervention on the other hand, Glen's comment about how the Mitchell River group was established illustrates how far some government representatives have to come before embracing the principle of self-determination:

> They had a tough time of it, trying to set themselves up while government agencies were kept out of it [and there were] internal conflicts. Now the group has made some progress over 12-18 months. They were missing the boat on a few things - it wasn't until we got involved and said "you can't take on so many issues - you've got to narrow it down". Then they became focused.

Talking about the role for government officers at stewardship group meetings, Glen also said, "you shouldn't lead them, but in initial meetings they might need leadership to get the group started". It wasn't clear whether he was suggesting that community groups need a facilitator to help determine their own goals and priorities (navigation), or was he inferring that groups cannot do anything without governmental assistance (domination)? Certainly, groups may need more government intervention at some times and less at others. Or, it may be government officers who need support and assistance from the community. Apart from the need for tact, there is no clear cut or simple technique for ascertaining what facilitation technique is appropriate. While mistrust between government agencies and community representatives concerning the freedom of any new stewardship group seems inevitable in the beginning, in the Mitchell River group's case there were signs that relationships built on trust and collaboration were developing among group members. Another government employee went as far as to suggest that the mere existence of the Mitchell River group had made his job easier and has provided "some great opportunities for interaction and liaison – it saves me a lot of stuffing around and gets people talking".

Facilitation, organisational culture and inter-governmental relations

A common theme among government departments is how other government departments (or sometimes different branches of the one department) are getting it wrong. It was not simply the age-old arguments between different tiers of government that were noticeable in the Mitchell River group's case (although these existed), it was ingrained conflicts between different State government departments. Blane's management of these tense inter-governmental relations was made even harder by a dual allegiance to the group and to the department funding his position. James, a senior bureaucrat in catchment management was surprised by Blane "operating cooperatively with other government people there – not confronting [them] in front of the group". This begs the question: how much confrontation went on between government departments behind closed doors?

Departmental culture also has a large influence upon the way in which a facilitator perceives his or her job. The corporate goals of the department, the office surroundings, the job opportunities available, the public service perks and day-to-day workplace culture all influence a facilitator's day to day work. Organisational hierarchy within government departments especially influences the style of facilitation. For example, facilitators must make quite considered decisions about seemingly insignificant choices such as whether to wear office clothes to work versus jeans; or whether to pack the jeans and change into them for field visits. All of these are critical in terms of whether s/he identifies him or herself as part of the organisational culture of an agency or as more fundamentally connected with the community. The consequences for facilitators making the "departmental professional" choice may lead to more responsibility and status conferred on him or her by the bureaucracy which "carries over into the community in the form of titles, prestige and so forth".[36] Yet the person still has to show his or her face in the post-office, bank, shops, school or sports facilities. So, if a facilitator misuses this power and prestige within rural communities to dominate groups, not only are they publicly scorned, but there is less opportunity for stewardship groups going down the path of self-determination. I did hear of such instances in other groups, but for the Water Watchers, Downside Landcare and Mitchell River groups, each facilitator was not coercive but animating, not powerful but motivating, and not overly concerned with the increased personal prestige of a 4WD vehicle complete with departmental logo. One final observation on the subject of facilitation in the Mitchell River group refers to the essential difference between community-based stewardship groups and government sponsored and established catchment committees:

> Look at whether a community group is really what you want. If you're trying to put pressure on other government departments, don't create a community group [to do this]. (Mitchell River government officer)

The extent to which stewardship groups are pawns caught in the competing strategies of rival government departments is seldom thought of as a problem. However, governmental cooption of groups, either in their emergence or establishment phase may be more of an issue. Although cooption was not a problem for Water Watchers, Downside Landcare or the Mitchell River groups, it is an important issue for groups who are reliant on government support. Stewardship groups are especially prone to cooption for three reasons:

> First, increased financial dependence on government can temper the prophetic passion of a voluntary organisation for fear of biting the hand that feeds them. Second, the style and objectives of an organisation can be modified excessively to serve the government's agenda. Third, the growing presence of voluntary association leaders in the governmental sector may subtly influence leaders of voluntary associations to be more cooperative with governmental agencies than they might otherwise be because of the lure of potential government service.[37]

Administrative workloads

> By law to get funding, we have to be an incorporated body and abide by the guidelines of the silly bureaucrats. A lot of paperwork is a waste of trees – it was quite difficult at first, this form, that slip, third party premiums, now you need an up-to-date form and it's a very difficult process. (Mike, Downside Landcare group member)

Community groups and government representatives alike believe that a central role of a facilitator is the shouldering of administrative loads. Creatively labelled "bureaucracy busting", effective administrative facilitation requires finding the appropriate forms; knowing how to fill them in and who to send them to; contacting the right person for advice and information within an organisation, and undertaking routine administrative functions (meetings, filing, bookwork). Increasingly, incorporation is necessary for stewardship groups so that individual members are protected against liability. Although many members perceive this as just another example of government red tape, it is necessary for insurance and regulatory purposes.

In terms of the administrative function they provide, facilitators interpret government guidelines, liaise with government departments and assist with the provision of technical information and specialist knowledge in a way that suits many stewardship groups. When talking about the administrative assistance provided locally by State government officers, Robbo of Water Watchers enthusiastically claimed that "the Community Catchment Centre is the greatest invention since sliced bread!".

Most interaction between members of Water Watchers and State government departments took place through the Community Catchment Centre (CCC). Government officers at the Catchment Centre were prepared to listen to local landholders and help them "fight their own battles" with State government using the facilities and administrative support provided. The responsiveness of officers working at the Centre to local peoples' concerns was especially noticeable. Water Watchers group members Jane and Lucy told me that "never once have we been knocked back by them". Another member spoke of the alacrity and reliability of response from the CCC: "if they say they're going to get back to you, they do". The breakdown of institutionalised structures was another point in favour of the Catchment Centre since it provided a base for officers from several different State government departments. Local residents talked of the building as a "one-stop shop". The resultant collaboration allowed a closer relationship to develop between the Catchment Centre, the Shire Council and State government services provided locally. According to Kevin who coordinated the Centre at the time, with improved public access to the CCC, more local self-reliance ensued.

Advising, training and disseminating information

> Some [government officers] are poor speakers – we want them to be able to tell us what to do in a practical way, and cheaply, but appropriately for the individual. Some government fellow told a woman [I know] all the expensive ways of fixing her property. (Ellen, Downside Landcare group member)

In this post-industrial era, exchange of information is as valuable a currency as cash. This is especially true at the community level where groups rely on government agencies and facilitators to provide technical expertise and professional advice. Government extension officers and district agronomists are still one of a rural community's best resources in Australia. Nowadays though, there is more emphasis placed on stewardship groups' access to those officers for specialist information and advice than there is on individual distribution of agricultural facts at field days and via technical memos and reports. This is quite different from the traditional role of extension scientists and departmental officers in the community. They are now reincarnated as educators, informers, advisers, and supporters in recognition of the fact that rural communities' information requirements are changing rapidly. In order to make sensible decisions based on local assessment, communities need both information and knowledge. As elegantly declared by an eminent academic friend, "data is not information; information is not knowledge; knowledge is not wisdom; and wisdom is not omniscience![38]

Members of stewardship groups frequently complain of information overload. However, although groups may be getting a lot of information, they may not be receiving the information they need in a form in which it can be usefully employed to make decisions. Traditional technical assistance and specialist knowledge has been provided to groups not just by government agencies but also by universities, private consultants and public education campaigns. Moreover it has been disbursed in a way which has historically discouraged individuals and groups from developing a local information system which meets their own needs. Wherever I went people commented on the inadequate provision of agricultural advice and information. This requires investigation:

> There has been misinformation given to the community – too many decisions have been made inconsistently as to what causes problems and what doesn't. (Robyn, Water Watchers group member)

In the case of Water Watchers, much of the (unrequested) information on nutrient enrichment provided to the Serpentine Jarrahdale community by government departments was not taken very seriously owing to the poor history of government-community relations. Conversely, the Community Catchment Centre was seen not only as an accessible source of technical advice, but as a place where landholders and other community members could exchange local knowledge and discuss common problems. The Catchment Centre was also regarded as a source of new perspectives on farming systems and eventually earned such a good reputation that landholders would seek advice from the "Catchment Centre blokes" before resorting to traditional agricultural extension services.

When members of the Water Watchers group requested information in the form of local environmental data on water quality from government agencies and departments outside their community, they often felt discouraged. Members of the group told me that they found the Water Authority of Western Australia was especially unhelpful regarding group efforts to compile local water quality data based on disaggregates of centrally held information. In order to get information from a government department by telephone, they said it was quite common to be handed from one person to another through three or four different sub-sections of the Authority before being finally connected to someone who may or may not be able to help. On a local or regional level, information collection and distribution is faster, cheaper and more efficient. Canadian research has confirmed that satisfaction levels with information provision are highest for local and regional government agencies and diminish in inverse proportion to the level of government.[39]

Clearly the role of information broker is another essential government contribution to stewardship groups. But it is more than just an economically effective way to get information out to the rural sector. Agricultural extension by way of field days, farm planning workshops or demonstration sites sponsored by landcare groups are central to a group's sense of empowerment and self-reliance.

> The evolution of landcare groups towards autonomy and self-reliance could be the basis for a move towards research, development, and advisory systems which are "bottom-up", "land user driven", "farmer first", rather than systems based on a communication model which assumes that most information

> flows in a linear, one way path from researcher, to extension agency, to progressive farmers, to the rest.[40]

However, there should not be an automatic assumption that this information is accessible to everyone just because stewardship groups are taking charge of their own science communication requirements. In Downside, Cathy would invite guest speakers and government specialists to general meetings and workshops. The professional was coming to the community, rather than the other way around, but this did not mean that the information espoused by that professional would be communicated in ways that allowed and encouraged widespread comprehension.

> Sometimes they talk a bit above you – I'd love to have been up the front and seen the blank faces. I don't think he could talk in more general terms. (Janine, Downside Landcare group member)

Some governmental representatives seemed acutely aware of this barrier to effective communication but appeared unable to resolve it. In their whole-hearted attempts not to indoctrinate the community with outside opinions and advice, I witnessed a few government extension officers who went to the other extreme and failed to provide new information or advice to community groups at all. Landcare group members are particularly aware of the removal of extension services from the past ratio of one-to-one advice, to the current situation in which an "expert" talks to a group of individuals. In some cases, the relationship of extension officer to landcare group can become very unbalanced in terms of power. When this happened in Downside, it served only to increase the divide between specialist agricultural knowledge and landholder practice.

While the setting for science communication may have changed from a one-to-one situation to a group setting, old habits of exclusionary language die hard among some extension officers. This is not so much a comment about the nature of the old guard, as it is about the need for more education and communication skills throughout rural institutions so that participatory practice in science communication is more widespread.

Another important trend in agricultural extension is a shift away from the model of "my cup of knowledge is full since I am an expert and your cup is empty, since you are the farmer, so if I tell you what I know then your cup will be full too". A new paradigm of extension science taught with an emphasis on participatory process is beginning to infuse current practice. It recognises the existing local knowledge of community members. It also legitimises the practice of sharing information and knowledge both among environmental stewards and with government agencies. This is quite different to the former model in that:

(a) it presupposes that landholders' cups of knowledge were not empty to begin with, there is a recognition of the importance of local ways of knowing;

(b) it recognises the importance of farmers sharing knowledge among themselves and not simply relying on one expert for direction and assistance; and

(c) it opens up important new avenues for local and traditional environmental knowledge to be taught by landholders and community residents to government officers and agencies.

Perhaps one of the most important roles of the Community Catchment Centre for Water Watchers was as a legitimate collection point for local history. Having established that the community catchment centre at Pinjarra was accessible and easy-going, it was not seen as simply a point source of government technical information, but also a place which respected local people's knowledge and especially Water Watchers' monitoring efforts. Rather than reject their data as non-scientific the Catchment Centre built their own monitoring profile in collaboration with the Water Watchers group such that both sets of data were complementary. This increasing recognition of the value of rural people's knowledge has enabled members of environmental groups such as Water Watchers to feel more comfortable in experimenting with local environmental stewardship.

During discussions with government representatives on the Mitchell River group, I noted some subtle differences among them. Although some officers believed that it was their role to give information to the group, most added that this was appropriate only when such information was requested by the group. Other extension services included providing advice on governmental policy or giving technical advice. In contrast James felt that the information service he was providing was from the community to politicians, via the bureaucracy, not the other way around. That there was not enough information collected or research conducted in the Mitchell River watershed was a frequently heard lament among stewardship group members. In other words, having identified a need for more two-way information exchange between government and the community, Mitchell River group members believed the catchment should serve as both a source of data and base for research not just a repository for government advice.

Environmental education

Water Watchers group members developed personal insights about water as a unifying feature of local catchments and appreciated the harmful effects of nutrient enrichment on the environment very acutely. This degree of "land literacy",[41] grew as government officers and members of Water Watchers collaboratively collected and analysed samples. One of the children who participated in the Water Watchers group said: "I thought it would be boring at first, but it wasn't". Another said: "We collected the water and Mr Brown sent it to the laboratory and had it tested. Before, we had the second highest reading - orange, now we have the second lowest - green" (red, orange, green and blue were useful symbolic indicators of the amount of phosphorus in the water). The proximity of the Catchment Centre to schools in the Serpentine-Jarrahdale area meant that project officers could talk to classes about the results of their catchment monitoring. The philosophical framework in which the centre operated also allowed adult members of Water Watchers to learn and exchange information, in their own environment and at their own pace. As Robbo said: "we planted conservation seeds in the mind and that's better than trees in the paddock".

Unfortunately, there is less emphasis on this type of skills-based project as a source of environmental education for adult members of stewardship groups. Despite the good intentions of many environmental education specialists and academics, it is publicly interpreted as something done *to* children, rather than a by-product of adult stewardship activity within a community context.

Expertise

With the advent of a new educative era that embraces the workshop, butcher's paper and team-building exercises, there is less reliance on "expert systems" for answers to environmental issues and questions posed by the community. The principle underlying these practices is important: government departments don't always know best and therefore shouldn't espouse their policies and approaches without consulting local opinion.

Information strategies which build on existing skills and local knowledge in the community are necessary. These strategies could encourage information partnerships between members and non-members of stewardship groups in an attempt to:

- find out the information and skills requirements of group members; and
- see if these can be met locally.

Stewardship groups need to know which educational strategies are likely to be effective with respect to local conditions. More management tools explaining the who, what, where and how of environmental care are necessary. For example, excellent material is produced by the Land Use Study Centre at the University of Southern Queensland. There are similar landcare manuals produced by the University of Queensland in conjunction with the Queensland Department of Primary Industries. The important feature of these resources is that *they can be appropriated by community groups to suit local needs*. If government departments and outside experts continue to act as if they know the realities of life on the land better than local landholders, mistakes based on lack of attention to rural people's knowledge will continue to occur. The quote by Salim, cited at the beginning of Chapter 1 is worthy of duplication here to illustrate how members of stewardship groups become angry at the thought that they have been duped into believing extension scientists and agricultural management consultants for years. This sugar cane and tobacco grower said:

> When I was younger – I always thought that these people in government departments could take care of it [the environment], but now I'm seeing that sometimes they get off the track ... We thought they would have had things more under control than that. I'm starting to realise that they're not much smarter than us – even with training, they still make mistakes.

Society can no longer solely and unquestioningly rely on extension strategies that are premised on positivist science and rationality, especially when attempting to redress environmental, economic and social pathologies in rural communities. Nor can governments afford to ignore local knowledge and choose not to listen to public opinion. Along with the realisation that governments do not always know best, the public are justifiably angered by the suspicion that community consultation means having to pay to have their ideas implemented, as I overheard in one stewardship group meeting. "When they [government] invite you to have a say, it means they want you to pay!". Although governments' attempts to really listen to community opinion and learn from local people are reassuring, at the same time there is deepening scepticism on behalf of some stewardship groups. Issues of consultation, participatory practice and democratic systems of representation in stewardship groups are also important features of government involvement in local environmental management.

Consultation

> Governments should not pre-empt or ignore what community groups want. [They should conduct] public participation and consultation processes. Governments should facilitate community groups' views ... Governments should listen to community groups, and be told to listen! Governments should not influence or tell community groups what to do. Governments should justify their decisions to groups. (Mick, Mitchell River government officer)

Needless to say, not all government representatives feel the same way as the above planner, who was well versed in the theory of consultation and participation. Many do, however, recognise that community groups have a legitimate role to play in collating public opinion, directing information to relevant government departments, and acting as information brokers from government departments to local residents. But on the whole, few government officers felt confident about the important practice of how, when and where to involve the public in consultation exercises. And worse than this, it seems that some feel quite threatened by stewardship groups who offer opinions and advice to government. As Greg said: "The trouble is the community is very divided, [there are] too many voices [and it] leads to great expense and no outcomes ... [there should be] no money from government where groups are acting as a consultation voice". Rather than engaging in consultation, this Mitchell River group officer believed that it was up to governments to set the agenda and have departments carry out their duties with the minimum of attention to local preferences. As an officer responsible for upholding government policy regarding land clearing and development initiatives during work hours, Greg felt overly compromised out-of-hours in his role as community member of a local fishing group. That group was attempting to halt development of foreshores on ecological grounds because of impending damage to mangroves and estuaries. He resigned from the fishing club on the grounds that "groups get in the way of development proposals – they're spoilers [and it leads to] time-wasting for governments ... [because they] bring undue pressure on developments without full knowledge or technical expertise".

Coming from the old school of "I'm from the government and I'm here to help you", Greg seemed unable to perceive stewardship group members as anything but passive recipients of government benefaction. He could not approve of dissension or diversity of opinion because, after all, he was the one imbued with legitimate government power to solve environmental problems. His view was that politicians develop government policy and his role was to implement that locally, with the minimum of fuss in order to get the job done. In one sense, Greg was right, governments are responsible for developing environmental policy even if others may disagree with the process of putting that policy into practice.

Putting public participation into practice poses difficulties. In a complex and turbulent planning context, with multi-party conflicts based on different values, ideologies and attitudes, various levels of expertise and different environmental settings, principles of good consultative practice have to be manipulated very carefully to suit local needs and conditions. Stewardship groups require new public participation methods tailored to suit local opportunities for negotiation, discussion, argumentation and mediation between government professionals and community groups. More specifically, they need opportunities for dialogue and exchange of perspectives, goals and methods. Formula-based methods of involving the public in governmental decision-making do not work. This begs a question about the kinds of democratic ideals in community groups which enable local residents to participate.

Participatory democracy

Government support for participatory democracy is well documented in theory but in practice it can be under-emphasised. Compare the case of the Serpentine Jarrahdale Land Conservation District Committee (LCDC) – the administrative body through which government funds were channelled to Water Watchers – with the group itself. The West Australian government endorsed public participation on such committees through official documentation and landcare plans. However, while LCDCs may encourage local participation on paper, they are anything but participatory in practice. The Serpentine Jarrahdale LCDC was comprised of two women (one of whom was the secretary) and fourteen men. Their meetings were run according to Roberts' Rules of Order, with statements typically preceded by "through you Mr Chairman". Four members of the LCDC were also members of Water Watchers. However, unlike Water Watchers' meetings, LCDC meetings invited government membership as technical advisers rather than facilitators. A noticeable feature of LCDC meetings was the preoccupation with power relations in the group.

> Now I feel that I'm the boss of the whole thing ... but what I've got to be very careful about is that we don't get lost in the people thing. It has to be [through the] Land Conservation District Committee all the time not a Community Catchment Centre thing. (Robbo, Water Watchers member)

The central point here is that the Land Conservation District Committee, although technically an environmental group, did not reflect the same easy going group style evidenced in Water Watchers and was not community initiated. It is therefore not a "community group" in the same sense as other stewardship groups and differed in several other important ways. Their group process was quite different. Power relations were more traditionally hierarchical than flat, probably as a result of the interpretation of the group's administrative function officially bestowed by the Commissioner of Soils in Western Australia. The *Decade of Landcare* document made this funding role quite explicit: "Increasingly, various grant schemes and incentives are tied to LCDC endorsement, in the knowledge that they have legal indemnity and are experienced in handling finances".[42] Although LCDC members included qualified accountants and administrators of funding schemes, they were not ardent advocates of community-based environmental management and were more likely to be there for political reasons than pragmatic ones. However, on another level, given the formal administrative functions of the LCDC, it may not be entirely fair to have expected them to behave any differently. Their formal style is quite fitting in the sense that they are the ones imbued with official powers to fulfil the administrative tasks set out by the State government.

Juxtaposed with Water Watchers' and the LCDC experience of power and politics, Mitchell River watershed group members had similarly political motivations for involvement, but operated quite differently. Members of this group were simultaneously the workers and the administrators, responsible for submissions, expenditure and projects, unlike Water Watchers who operated under the administrative protection of the LCDC. However, like the Land Conservation District Committee, members of the Mitchell River group possessed similar motivations for participating in the group; for political prestige, and to make sure that the "right" industry sectors and geographic regions of the catchment had a say. This representational membership system was significant in terms of the way it increased the likelihood of conflict within the group.

Representation

Representation, when broken into its composite syntax, means presenting again or re-presenting. In practice, representation often means delegation: one member represents an amalgamation of views, attitudes, beliefs and ideologies of non-members. For example in the Mitchell River watershed group, two government delegates represented the Department of Environment and Heritage, and two others represented the Department of Primary Industries.

On catchment management committees, interests are frequently divided into sectoral or social categorisations such as graziers, miners, Aboriginal people, and fishers. But that division is not only simplistic, it ignores the interests of future residents, those who have no voice or those who are less capable of representing themselves. Surely they also have a stake. Other forms of participatory resource management acknowledge local interests on international livestock corporations for example, and custodial rights in mining company boardrooms, where decisions are made which affect the lives of those living in the region. Aboriginal members of the Mitchell River group, while officially represented alongside graziers, miners and government employees have a different approach to decision-making and quite different relationships of power to others represented. Their interests and decision-making processes cannot be divorced from the broader context of their land, their country and their kin. Nor can their participation (for the majority on low incomes) be considered equal to those who earn four or five or ten times as much money per annum. As for fishers in the Mitchell River group, while represented on paper, they are rarely able to attend meetings which have been called outside the times and places generally accessible to that constituency.

In other words, not all those represented in equal numbers or proportions can make equal representations to catchment management committees:

> A fair negotiation process cannot be assumed between unequal members and compromise may be elicited from weaker landscape proponents where none should have been required.[43]

Representational systems of democracy, without the commitment to community-based planning ideals, have distinct limitations in terms of environmental stewardship. Committee systems of decision-making set up by governments to represent all the views on environment and development within a specific catchment are often well conceived on paper, but lack a sense of territoriality without which they also lack the conviction necessary to fulfil long-term goals. They do not represent all those whose interests are already known let alone those whose interests are still unstated in their quest to represent the major industry sectors and the more populated localities. Elderly people, children, ethnic minorities, Aboriginal people and women are among those whose perspectives are frequently overlooked or assumed by others. As Blane said, "greater female representation is needed [on the Mitchell River Watershed group], especially regarding their traditional concerns of women. No one speaks of the quality of education, I'd like to see that reflected in the committee [structure]". Although representing other people's issues is widespread practice on ICM committees, it is ultimately flawed, since one person's experiences cannot be objectively recapitulated by someone else "as they really are", independent of the experiences and beliefs of both represented and representer.

Representational membership systems breed conflict within groups since the role of the representer is to bring the demands and views of their constituency to the negotiating table. In taking on the position and demands of an entire organisation, delegates defend both their organisation and its interests thereby further diminishing the opportunity to calmly investigate all the issues and create alternative options. In the case of the Mitchell River group, there were several occasions early in the establishment phase of the group where conflict between members taking adversarial positions threatened to disunite the group. As long as individual members represented an organisation it was apparent that they were not engaging in the issues and interests of other members. While representing someone else, members cannot engage in debate and discussion from their own personal perspective, which ultimately will damage the potential to move toward a more open system of communication. Or, if they do engage in debate on a personal level, delegates can be criticised for not representing their organisations accurately. Having said that, representative systems of decision-making are still the most appropriate for large-scale ecologically sustainable development within complex and uncertain socio-economic and bio-physical environments.

Conflict generally leads to more entrenched positions and less ability to investigate common solutions to problems. In the Mitchell River group, one of the representatives from the Department of Environment and Heritage (Leanne) had responsibility for fire management within the catchment. That department has a history of policies which conflict with traditional practices of many Aboriginal communities. Before coming to the group, Leanne envisaged a conflict with the Kowanyama Aboriginal community about fire regimes. However, through open discussion and exchange in informal settings, that potential conflict did not erupt, and is unlikely as long as there remains:

> [a]n awareness of diversity, the politics of difference, the fundamental claim of otherness which all point to a politics based on democratic debate and decision-making in an atmosphere of mutual tolerance and respect for differences.[44]

Regulation

Regulation is still perceived by rural people as government's traditional role in remote areas. Government officers are still associated with stock control, monitoring of mining leases and generally acting as "land and environment police". To a large extent the historical legacy of government control has carried over into the current era of more open and consultative government practice. Nowadays the degree of control acutely depends on whether government has a statutory or common law responsibility for a specific resource management issue. Despite this, an underlying mistrust of government resides not far beneath the surface of rural communities. For the Mitchell River group, one government officer advised others to "tread cautiously - be aware of the fear of government - be sensitive [to community needs]". Glen suggested that catchment management committees should not "let too many government people [on them], otherwise people don't want to have anything to do with it". Patrick and Leanne were similarly cautious:

> Government have a duty to see the health of the river lifted – but we can't do it with a big stick. (Patrick, Mitchell River group government officer)
>
> Wear your uniform, but leave your big boots at home. (Leanne Mitchell River group government officer)

The perception that government officers are there to see that regulation is enforced is still a reality in the minds of both government officers and community members. However, both governments and stewardship groups now advocate a softer management style which favours consultation and public participation. In reality, regulation as a function of government involvement in community groups is low. Regulation is associated more with individual properties and specific landholders than any group. However there is a perception by stewardship groups that the regulatory function of government still exists – only these days it is disguised as monitoring and evaluation of group practice.

Evaluation

Evaluation is a complex topic with many nuances and subtleties of interpretation. Here I wish to provide a brief overview of evaluation as it pertains to the increasing role for government involvement in stewardship group activity. There are two main questions which most current evaluations are attempting to answer:

- Where and what is the on-ground change resulting from this project?
- Where has the money been spent on this project?

A third, perhaps more important question is often ignored in evaluations of stewardship group activity: where is the national political action required to address the chronic problems afflicting our environments?

Evaluation of stewardship group efforts, given the nature and diversity of the projects themselves, is a little like measuring a charging elephant with a rubber yardstick, to cite an evaluator's quip. There are many questions still left unanswered: Who is the evaluator? What are their goals? How do you take each group's objectives into account across a wide range of differing projects?

How can the evaluation be objective? Is that desirable anyway? What kind of measurements can be made? And how do you recognise success when you see it? Is it limited to those groups who have fulfilled their objectives? And so on. Although it is pertinent that these questions be raised, the issue of whether and when they can be addressed is both politically and technically problematic. This difficulty is exacerbated by the absence of an established database of indicators of ecological sustainability (despite the preliminary attempts illustrated by the series of Australian State of the Environment Reports). Evaluations of stewardship group activity are less than meaningful without a baseline with which to compare groups' results. Another difficulty is the lack of interdisciplinary thinking in this field. It is commonly assumed that landcare group's successes, for example, can be measured in terms of the extent of biophysical change in a catchment area; more trees, less salinity, less vertebrate pests. But a stewardship group's success may equally be measured in terms of the nature and extent of participation, negotiation, collaboration, environmental education, and so on. Questions remain unanswered about the interaction of socio-economic and cultural goals and indicators with bio-physical ones - assuming there is a common paradigm and set of methodological tools within which to locate such an analysis!

Answers to some of these questions can be found in the analysis of differences in approach to evaluation. Many evaluations are conducted at the end of a program's life by an "independent" evaluator for accountability purposes only. They do not purport to find out why a group's actions failed, or why objectives were or were not met, or who was involved at various stages in the project. Other evaluation styles attempt to involve a range of stakeholders in the evaluation, to plan the evaluation process early in the project's planning or implementation stages, and to establish how the evaluation will influence future project activities.

Another category of answers to evaluation questions within the context of environmental stewardship is to be found in the differences between group members' perspectives and that of others involved. From a member's perspective, success is closely linked to the extent to which a specific project's objectives are met. After that the group may split up and discontinue its efforts. Or, members may remain in a group and continue to act incrementally making very slow progress on the ground whilst consolidating group cohesion. The latter is often defined by both community and government as "failure". But from a group member's perspective, failure may amount to their inability to affect government or corporate policy, process or planning around a specific environmental topic. In other words, they have expectations of what their on-ground actions may lead to. Some are unrealistically high perhaps, but the point is that the evaluation is not all one-sided. Governments are often put under the microscope by groups in terms of their willingness to act, to listen and to appreciate stewardship groups' efforts.

From a governmental or corporate perspective, accountability for monies spent is extremely important in any evaluation. Length of fence lines established or number of trees planted become the yardsticks of success. However this preoccupation with things numeric cannot begin to understand the complexities of evaluating group process, program feasibility, needs or priorities. Funding agencies are frequently disappointed with the apparent inadequacy of stewardship groups' efforts to make a difference to local environments. But landscape changes cannot be easily or quickly determined. Similarly social "impacts" or changes cannot be comprehensively understood in snap-shots. Hence accountability factors are frequently used by evaluators to the detriment of incremental changes in stewardship approach and landscape processes.

Many stewardship groups are now implementing environmental monitoring projects. For example they may be establishing the extent of salinity-affected land, finding out about the habitat requirements of platypuses in agricultural areas or recording water quality data. There is a resounding need to promote discussion between agency and community scientists in terms of the relationship between monitoring and evaluation, establishing who gets to judge, and deciding on the links between monitoring, evaluation and decision-making.[45] Scale issues and political issues abound in terms of equity between groups and assessment of the overall impact of any program intervention. There must be a balance of efficiency questions as well as the more quantitative effectiveness and accountability criteria in any evaluation. Without this, the continued funding of stewardship projects can be interpreted as a political act of faith or a foolish waste of public money, given the notoriously short political time frames and the climate of economic rationalism which requires a return on investment.

Whatever evaluatory and regulatory responses are made to environmental problems, they cannot be implemented across the board without reference to local conditions. The old command and control style of public administration should be, and is being, phased out. More flexible approaches to environmental management which acknowledge the central role of voluntary conservation and stewardship initiatives are required.

Summary

I do not share the populist but naive expectations that stewardship groups can be empowered to shoulder all local environmental responsibility. Community action cannot proceed outside or against powerful State or private institutions. Local community groups are not romantic back-to-the-landers, nor hard-working rural folk united in their common goals, sharing common ideals and working together toward a common future. Similarly, stewardship group action is not independent of the political structures and processes of other democracies. "Although an alternative development must begin locally, it cannot end there. Like it or not, the state continues to be a major player".[46]

This chapter establishes that it is very important to investigate the role of government in helping or hindering community groups. Behind the current push for more information about community groups lies a serious concern over how to implement environmental management policies in regional areas. Of the five principles discussed in this chapter – *resourcing, facilitating, informing, consulting and regulating/evaluating* – much of the discussion has focused on the moral persuasion end of government policy, the carrot as opposed to the stick. This approach has been labelled the "social response" model, set up by governments in an attempt to have communities respond to local threats or opportunities. The goal of this model is "to shape public attitudes through normative persuasion rather than prescriptive direction".[47] However, that presumes that there is some shared vision of ecological sustainability between communities and various levels of government. What is reflected in the three stewardship groups is somewhat different in that there is seldom such agreement across government departments let alone between government and community groups. And there is often outright disagreement. It is foolish to expect that there would not be competing interests, values and beliefs across government departments as well as within and between such diverse communities as the Mitchell River watershed and the Downside Landcare group.

These principles of government involvement provide a structure that enables assessment of the relative importance of each principle to the cooperative management of local environments (see Table 5.4). The next chapter investigates principles of environmental stewardship from group members' perspectives.

Table 5.4: Principles of government support for local environmental management across three stewardship groups

Principles	**WW**	**DL**	**MR**
Resourcing			
— in kind	⬤	●	•
— submissions/timing	⬤	⬤	•
— value-added accounting	●	●	•
— rural-urban divide	●	•	⬤
— resource use	●	●	●
Facilitating			
— facilitator-group relations	⬤	⬤	●
— style of facilitation	⬤	⬤	⬤
— organisation culture	●	⬤	●
— administration	●	●	●
Advising, training and informing	⬤	⬤	•
— environmental education	•	●	•
Consultation and participation			
— consultation	•	•	●
— participatory democracy	•	●	⬤
Regulation and evaluation			
— regulation	•	•	⬤
— evaluation	•	•	•

Key: • Minor importance ● Some ⬤ Major importance

Endnotes

1 Where policy is defined as: "a statement of intention; a definition of how things should be and an avowal that the institution in question will do something about it" (Dovers 1994: 1).

2 These were prepared by the Department of Home Affairs and the Environment 1984, the Commonwealth of Australia 1992 and the Council of Australian Governments 1992 respectively.

3 Pfueller (1995: 28) writes convincingly about Australian policy context on water quality programs.

4 For an extensive examination of matters of environmental policy, sustainability and uncertainty, see Dovers 1995; Dovers and Handmer 1995; Hamilton, C. and Throsby, D. (eds.) 1998.

5 Dovers and Lindenmeyer 1997: 73.

6 Blaikie et al (1997) made distinctions between classic, neo-liberal and neo-populist paradigms of environment and development to emphasise how and where local participants' knowledge contributes to research.

7 Partnership agreement – Natural Heritage Trust 1998.

8 Although it appears as if there are more organisations than projects, this information also relates to non NHT programs and projects. Additionally, many duplications arise given slight spelling variations, hence two records are created for the same organisation. More accurate NHT reports and databases are still under construction by officers at Environment Australia.

9 See the Australian Bureau of Statistics 4602 publication series *Environmental Issues: People's Views and Practices.* In 1999, 69% of people reported having environmental concerns, compared with 71% in 1998, 68% in 1996, 69% in 1994, and 75% in 1992 (*http://www.abs.gov.au/ausstats/abs@.nsf/Lookup/NT00012F3E*). See also Lothian 1994; Crook and Pakulski 1995; Pakulski, Tranter and Crook 1998.

10 Dovers and Lindenmayer 1997: 67.

11 The Save the Bush Tool Kit was developed by Charles Sturt University and released in July 1997.

12 Dovers and Lindenmayer 1997: 76.

13 This is an indication of the "outsourced environment" according to Libby Robin (pers. comm. September 2000).

14 Hollick and Mitchell 1991.

15 Campbell 1992a: i.

16 After all, many were offered seed money by government and were not entirely motivated by the grass-roots activism that captured the peace and women's movements.

17 Rick Farley's speech notes to the National Press Club in Farley and Toyne, 26 July 2000.

18 Government of Victoria 1988.

19 Hollick and Mitchell 1991: 2.
20 Burton 1992: 8.
21 For more detail see Beilin 2001: 6 and Beilin 2000.
22 Beilin 2001: 12.
23 This reflects the involvement of Terry White, (then employed by VSB) in helping establish the group.
24 Bradby 1992: 144.
25 There was one notable exception to this exodus. Kevin, one of the senior WADA officers who had helped set up the group was very much an advocate of local people and local places. He lived in the area and with his distinctive and forthright personality let everyone know what it meant to be working *for* the region and living *in* the region.
26 An indigenous land management agency based in Kowanyama.
27 Millar 1994: 9.
28 See Osborne and Gaebler 1992 for an overview of the international arena in new public managerialism and Curtin 1999 for an excellent appraisal of more recent Australian examples of civic discontent.
29 Grabosky 1994.
30 Grabosky and Gant 1999: 151.
31 Campbell 1992b: 167.
32 Bradby 1992: 146.
33 Warner 1989: 118.
34 Although he went on to acquire skills in these areas.
35 Littrell and Hobbs 1989: 56.
36 Warner 1989: 127.
37 Langton 1982: 9.
38 Henry Nix, CRES, *pers. comm.*, 2000.
39 Gardner 1991a.
40 Campbell and Junor 1992: 16. See also Pretty and Chambers 1994 or Scoones and Thompson 1994 for more insights into "farmer-first" and "beyond farmer-first" philosophy.
41 See White 1992; Mitchell and Stapp 1995 and more recent examples taken from Alexandra, Haffenden and White's 1996 publication *Listening to the land.*
42 Department of Agriculture 1991.
43 Beilin 2001: 9.
44 McEachern 1993: 35.
45 See Mahanty, S., Guijt, I. and Carr, A. 1997. (comp). *Tracking change: community monitoring in natural resource management*, proceedings of a workshop facilitated by Anna Carr and Irene Guijt at the Australian National University in June 1997 for more detail on this topic.
46 Freidmann 1992: 7.
47 Synnott 1991: 43.

6 Principles of community involvement

So far I have described how stewardship groups operate and how governments assist these groups. This chapter draws out the principles by which community groups work, investigates them through detailed examples and links them to theory. These principles are not recipes, which necessarily apply to all stewardship groups. Rather, the three stewardship groups provide the basis for generating principles, which may be considered in relation to the conditions necessary for other stewardship groups to care for their environments.

This chapter has three stages. First, Table 6.1 presents each of the substantive principles from all three stewardship groups, according to their relative importance in each case. Second, each principle is discussed, which comprises the main part of this chapter. The positive features of each principle are emphasised, but attention is also drawn to negative aspects to illustrate the complexity of group dynamics. Again, it should be reiterated that there are no simple rules, which if followed would ensure efficient and effective environmental stewardship. Third, the principles identified are analysed according to their relevance to each group.

I found four key principles for community-based environmental stewardship common to each of the groups I visited: a sense of place, a sense of community, a local knowledge system and empowerment. These principles will be discussed in turn.

Sense of place

The definition of place used here originates from Ted Relph in his astute and sensitive treatise, *Place and placelessness*. It is long, a generation old, but still very pertinent.

> Places are fusions of human and natural order and are the significant centres of our immediate experiences of the world. They are defined less by unique locations, landscape, and communities than by the focusing of experiences and intentions onto particular settings. Places are not abstractions or concepts, but are directly experienced phenomena of the lived-world and

> hence are full with meanings, with real objects, and with ongoing activities. They are important sources of individual and communal identity, and are often profound centres of human existence to which people have deep emotional and psychological ties.[1]

Each group operated within a defined place. Water Watchers used the administrative boundaries coinciding with those of local government. Being on the outskirts of a metropolitan area, many members described their attachment to land in both rural and urban terms. The group's boundary was complemented on a small scale by each member of the group taking responsibility for monitoring a creek or drain near where they lived. Thus members spoke of Robyn's area instead of the Oakford subcatchment, or about one of Robbo's sites, instead of Beenyp Brook. In this way group members attained a sense of biophysical or catchment boundaries in relation to their houses and local surroundings, as well as acknowledging the standard administrative area bounded by the Serpentine Jarrahdale Shire.

Table 6.1 Principles of community involvement in environmental stewardship for Water Watchers, Downside Landcare and Mitchell River groups

Principles	**WW**	**DL**	**MR**
Sense of place			
history and attachment	•	●	●
boundaries and territory	•	●	●
collective identity	•	■	•
Sense of community			
membership and belonging	■	●	■
reinforcement	•	■	•
influence	•	●	■
emotional attachment	•	●	■
Local knowledge			
awareness and recognition	■	•	•
ways of knowing	■	■	■
local and traditional knowledge	•	•	●
Empowerment			
equity	■	■	•
self-determination	●	■	●

Key: WW = Water Watchers, DL = Downside Landcare, MR = Mitchell River
Relative importance • = Minor ■ = Some ● = Major

The Downside Landcare group defined its sense of place in terms of the district in which everyone lived. Cleared paddocks and rolling hills provided the backdrop against which a uniformly rural environment was divided by two valleys. These valleys ostensibly provided the geographic catchments from which group membership was drawn. However, community ties and the cultural geography of the area were, in practice, far more important factors in defining Downside's sense of place.

The Mitchell River watershed group drew a boundary around their catchment following the ridgelines and hilltops. However, it was derived primarily from digitised surveys of spot heights and stream flows given the apparent flatness of the landscape. With the diversity of membership and conflicts in interests, there was little within the largely invisible catchment boundaries to band the group together. Members of the group stated that they were moving toward a more bioregionally defined sense of place, putting more emphasis on the role of the river in uniting communities at the top and bottom of the watershed. However, each member organisation or individual also acted independently toward its protection. The importance of the river seems most obvious for Aboriginal residents in Kowanyama at the bottom of the catchment near where the Mitchell River flows into the Gulf of Carpentaria. They have defined themselves in relation to that landscape for many thousands of years.

History of and attachment to place

Attachment to place is culturally and socially defined. Just as one's own perceptions of place are dependent upon one's experience and memories, so attachment to place is influenced by our culture. Over time, a community's attachment to place increases, especially over a number of generations and with cultural artefacts marking inhabitation of that place. Relph states that "much ritual and custom and myth has the incidental if not deliberate effect of strengthening attachment to place by reaffirming not only the sanctity and unchanging significance of it, but also the enduring relationships between a people and their place". He maintains that when people lose contact with the myth and rituals of their places the places themselves become "changeable and ephemeral".[2]

The customs and rituals which sustained the Kowanyama Aboriginal Community's attachment to the floodplain at the west end of the Mitchell River watershed were forcibly weakened when "development" projects interfered to displace the myth and sacred nature of "story sites" or "story grounds". Aboriginal members of the Mitchell River group told me of their fears for the future of the river and their anger at not being consulted over decisions, which affect their lives and the life of the river system. Charlie spoke about the destruction of a rare lotus bed when a lagoon was dammed and drained to make way for the construction for a new road.

> The Department of Community Services constructed a dam at Red Lily Lagoon ... but didn't ask us! They buggered up the whole place. They should have asked the elders for that place, Kokoberra people. That [was] a story place and now [there are] no more red lotus lilies. They also buggered up Sandy Hole lagoon, trying to dam it up. That place was a story ground too. That lagoon was there since [the] very beginning of creation and was looked after by those Kokoberra people. (Charlie, Mitchell River group member)

When places are destroyed, so ritual and myth lose their significance and attachment to place declines, but the reverse is also true. Through historical attachment to place, rituals are created and an enhanced sense of place can eventuate. An English Royal Commission on Local Government in the 1970s found that attachment to home area increased with the amount of time that people spent there, an observation equally applicable to stewardship groups. Out of the 48 members of the three stewardship groups, 21 had moved into the area 21 or more years ago. Of these, 16 were from the Downside Landcare group, a disproportionately high number compared with the Mitchell River and Water Watchers groups. Aboriginal members of the Mitchell River group were also long-term residents but they were very much in the minority in terms of the membership base of that group. In the Downside Landcare group, sense of place was particularly significant as an identifying feature of the group. It distinguished members and non-members along territorial lines, and illustrates that attachment to place grows with length of time spent in the area. Sense of place was strengthened in Downside through landcare members' feelings of identity with and commitment to people sharing a common landscape.

These feelings of commitment and identity are shared not just with other members of the group, but are perceived as attachments to the very ground itself. This concept of rootedness goes beyond simply "caring" for a place to also imply a real responsibility and respect, hence "stewardship". In addition to society's general need for protected spaces, specific people have a need to belong to specific places because of the culturally constructed nature of these geographical and social communities.

The sense of camaraderie displayed by members of the Downside Landcare group united people attending landcare group meetings. This is linked to the presence of the tennis club in the centre of the region. Meetings are held there, the seedling nursery is there, people treat it much like a community hall. Everyone knows where the key is kept and most play a role in maintaining the building.

> The tennis club is our main hub around here, [it's our] main focus for bringing people together [for the] social side and exchanging ideas and experiences. (John, Downside Landcare group member)

> The tennis club is a pretty big draw card for the district; it's a good community, whatever is going on, people get involved. People are generally interested in the whole of the district. (Leo, Downside Landcare group member)

Because the Downside Landcare group increased members' sense of community, members were less reliant upon Wagga Wagga, a large regional centre nearby. When talking about the group, John said: "it keeps people together – focused on this community, not town". And Daphne lamented days gone by when Downside, and the even smaller Coursing Park, were recognised communities in their own right:

> It is something that has changed with transport et cetera … In the early days this was a real community, tennis club, church and hall, but now with farms getting larger, less people here and more transport [into Wagga Wagga], so there are less district community groups.

Emotional attachment to place was evident in each of the stewardship groups. However, there was a marked difference between those groups who defined their relationship with the environment in terms of territory and those who classified it as a functional relationship.

Boundaries as territory or function

Here, I am not using territory in its ecological sense and do not wish to draw any analogies with the way that other sentient beings claim territory. I prefer the term territorial as opposed to territoriality, to define the organisation of space from an insider's perspective, rather than an outsider's perspective.

Water Watchers

Water Watchers displayed territorial notions of attachment to place when they expressed a desire to "put something back into the neighbourhood". Jim told stories of his childhood in the area, about the number of birds and animals he would encounter then. Ilsa's family had originally come from this area and as a girl she had grown up in the Shire. Both people had moved away, but now felt some sense of affinity to the land and a desire to protect it.

Despite these territorial ideals, Water Watchers has a largely functional relationship with its environment. Members were interested in phosphate levels from the perspective of proving to the State government scientists that phosphate run-off was not emanating from the farms or other properties at the top of the catchment. They had a subsidiary interest in planning changes to their agricultural practices, but did not have the strength in unity or commitment to put their plans into action. Collectively they were not driven from a territorial relationship with their common property. Instead, some members clearly perceived a division between those members who lived "up the hill" and those who lived "on the flats".

Downside Landcare

The experience of Water Watchers is in marked contrast to the relationship of the Downside Landcare group to their environment. The latter was founded almost purely on territorial affiliation to a common district between farmers in two adjacent valleys. Families in Downside who had been there for nearly ninety years formed a close bond with the landscape which identifies their place or property in relation to the local geography. Social boundaries are very much a reality in terms of the formation of the landcare group. Forty to fifty years ago, people who "lived over the hill" from one another may never have met, due to the social and geographical patterns of where they shopped, played sport or went to church. With school closures and greater ease of transport, residents of the whole district now band together to create a sense of community, or risk being socially isolated and peripherally dependent on the urban centre of Wagga Wagga.

During my visits to Downside, a group member would quite often point up or down hill to locate his or her property in relation to others in the valley. This is meaningful considering the effects of salinity affecting individual properties. The geographic separation of recharge and discharge sites was perceived as a problem for some group members who understood that salinity itself recognises no social or administrative boundaries. Ecologically and economically damaging processes may affect a region some distance away. Territory then becomes less of an individual issue and more about identification with a catchment, or community group. Rhiannon said:

> [With] a community group, everybody has self-interest and if they have a problem they don't all have the time or money to fix individuals' problems but as a group they can help each other and in the process help the whole district.

Mitchell River

The Mitchell River group had both a territorial and functional relationship to their environment, recognising the unique cultural, historical and social characteristics of the area as well as its ecological and physical dimensions. This sense of belonging to the north of Australia characterised the group and was echoed through the sentiment "we're different from the south, we don't have the same values, beliefs and attitudes as southerners do". Another example of the Mitchell River group's territorial attachment to their catchment was the often-expressed desire to protect the Mitchell River, not to allow it to become sullied and dysfunctional, as members perceive other (southern) river systems have become.

The territorial relationship with the watershed was also evident when people insisted that they were still living on the edge of society. People saw Cape York as the last frontier. But perhaps what most characterised that group was the diversity in ways of seeing the landscape. Graziers, miners, tourists, fishers and Aboriginal people all had different ways of valuing the catchment. Although the overt goals and objectives for the watershed were stewardly, the covert reasons for preserving their catchment were a slanted version of the tragedy of the commons based on members' private attachments to public lands. Some were interested in environmental management of the Mitchell River watershed for personal gain, rather than its ecological integrity.

Collective identity

Despite a legacy of building fences and operating within a culture that idolised the rugged individualist, these community groups have collective experiences of district pride. In the case of Downside, field days for tree-planting projects on local farms served to bring private property into the realm of public concern, restoring a collective identity to the community.

> In short, people are their place and a place is its people, and however readily these may be separated in conceptual terms, in experience they are not easily differentiated. In this context places are "public". they are created and known through common experiences and involvement in common symbols and meanings.[3]

Signs erected on demonstration sites are common symbols in Downside and unite the community in two ways. Not only do they interpret the landcare message for passers by, they also serve to reinforce common boundaries and contribute to the construction of collective identity. But it is not only the identity *of* a place such as Downside which is important to group members, a feeling of identifying *with* a place was also important, particularly in terms of being insiders (members) or outsiders (non-members). For group members, there was an even finer distinction between general group members and committee members. Figure 6.1 illustrates the proximity of members' and committee members' properties at the southern end of the district.

Landcare group members' identity with the district was strengthened by local historical knowledge. The original chairperson of the group, Mike lives on "Narua". a property near the middle of the catchment. He also owns "Coursing Park", one of the first settled properties in the district and owned by his father before him. An elderly neighbour explained the strong ties between these two properties and the original public house (or pub), whose proprietor had named the district after an English country town. Although they may seem out of place in such a typically Australian landscape, Anglophile property names are a widespread phenomenon across the continent.[4]

Figure 6.1: Downside Landcare group – property relationships

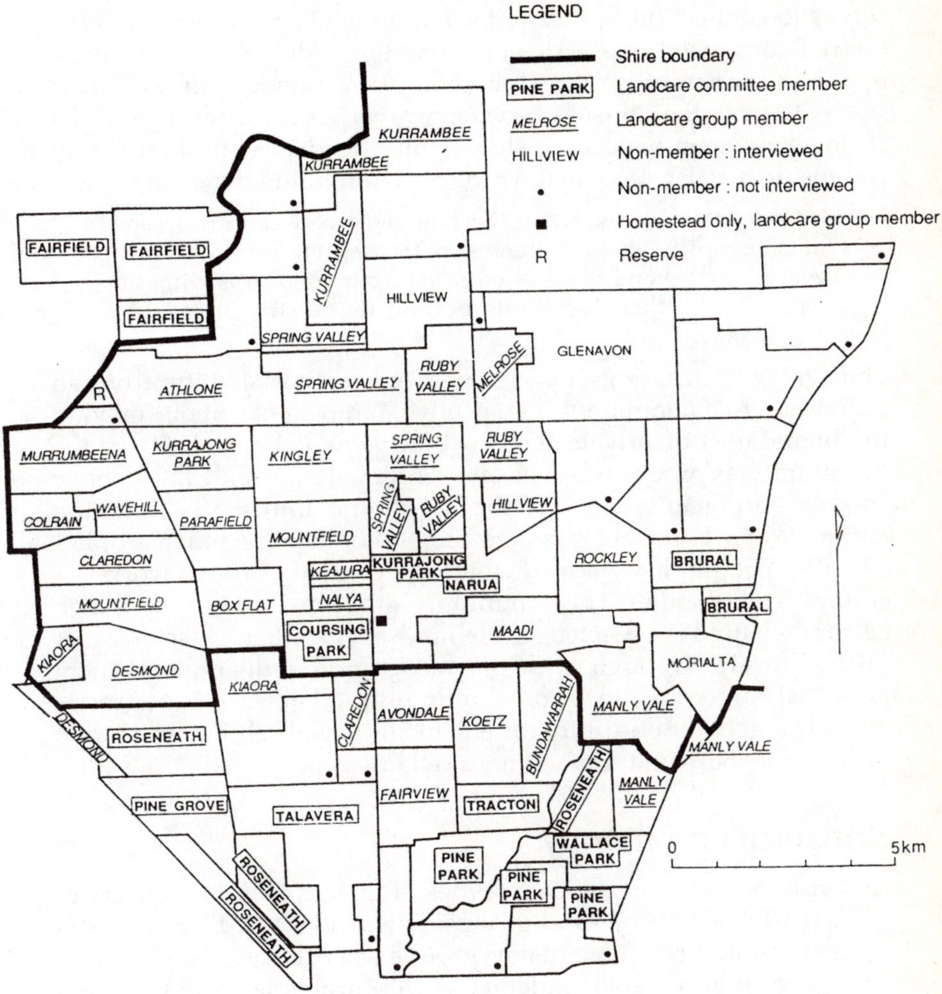

Manifestations of insider and outsider identifications are never more apparent than at junctions, gateways and thresholds. To strengthen common bonds, members of the Downside Landcare group display their membership status on the front gate to their property with a metal sign complete with landcare logo. This sign serves to connect the spaces of landcare members. No longer is the district comprised of a series of separate inside spaces, it is made up of an apparently unified boundary of "landcaredness". This serves to push the division between properties aside, in favour of a stronger division between insiders and outsiders within the larger Downside district. One final word from Relph illustrates this point.

> As our intentions vary, so the boundary between inside and outside moves ... if our interest is focused on our home, then everything beyond home is outside, if our concern is with our local district, then everything beyond that district is outside, and so on.[5]

The Water Watchers group also showed territorial shifts from an individual to a community perspective. Jenny spoke about moving the boundaries of private responsibility when she said: "I see the community as an extension of my garden". Janine and Lucy spoke about wanting to get out of the house and into the community; joining Water Watchers was their method of meeting other people.

The geographic place in which stewardship groups work is a cultural construction. The community group is a mirror by which cultural change is reflected in the landscape, either consciously or unconsciously. Through conscious recognition of the reinforcement of social processes in rural communities, stewardship groups provide a good opportunity for enhancing social self-determination in changing both cultural and natural landscapes.

Sense of community

A sense of community contributes a great deal to concerted, participative action by local groups. It was identified by members of each of the three stewardship groups as critical to their success. Therefore, it is vital to understand this principle in detail, both theoretically and from what people told me about it. Sense of community arises out of the fulfilment of four distinct criteria: membership, influence, emotional connection and reinforcement of needs. These form the basis of discussion introduced in Chapter 2 and are now extended.

A sense of belonging to the group in a local area is frequently given as the reason people join community groups. For members of Water Watchers, simply "belonging" was considered a more important reason for joining than altruism, personal gain or meeting new people. The concepts of belonging and exclusion were operating very clearly for children at the Serpentine Jarrahdale School during a classroom discussion I took part in. They told me that two key factors that stop people joining community groups were "people tease them [about] joining groups" and "not having friends in the same group". Adult non-members of Water Watchers echoed these comments. Their wish to avoid the tag "radical greenie fringe" sounded a lot like fear of being teased. The importance of participation and avoidance of exclusion were noticeable features of all discussions with members of Water Watchers.

In the case of the Downside Landcare group too, there was an obvious relationship between members' sense of belonging to and identification with groups and enhanced participation in community activities. In studies of landcare, belonging, involvement and participation are cited as key elements of a successful group. Many of the Downside Landcare members spoke about how highly they valued their involvement with the group. In Downside there was a high participation rate for women (discussed below) and for people of different ages. Across the whole Downside community, belonging to and identification with the landcare group increased the ownership of the landcare ethic.

Research into who participates in voluntary community organisations raises some questions of relevance to community involvement in landcare. It is confirmed that people who strongly identify themselves as part of their community are more likely to get involved with local organisations. This implies a need for more research into the *nature* of belonging and not simply the *numbers* of participants, since the quality of participation is related to people's sense of community. Also, the link between belonging to a community group and participation in community activities is bi-directional. The converse is also true. The more people identify with a voluntary association, the more they participate in broader community activities.

However, there is a cautionary footnote to this happy-ending story of membership and belonging. A source of discomfort for Downside members was that there were still some landholders who were not members of the landcare group. Members wrote about some of their feelings of dissatisfaction:

> It's hard to get everyone keen and participating.
>
> Not being able to get neighbours involved where the recharge area starts [was a negative experience].

Non-members, on the other hand, typically reported that the landcare group was doing well and that they (non-members) had a good relationship with the group, from a distance. That was a literal distance in several cases. People living on the boundaries of the landcare area were generally less involved than those living close to the Coursing Park Tennis Club. Other people who lived outside the area also spoke of being "sympathetic to their cause ... [but] haven't had much direct involvement". Other non-members of the group felt that they had to justify their non-involvement.

> [I have] a good relationship, [they're] good neighbours and good chaps, good family people, I couldn't say one word against them. I've got on well with them. I watch what they're doing, but don't agree with spraying.[6] (Raymond, Downside)

Another non-member spoke about not belonging in terms of a fundamental disagreement with the way the group was run.

> In the last 12 months, I haven't had anything to do with it, not for any personal reason, just because I'm against the way it's operating. They got a grant for $23,000 for mapping. When the mapping was introduced, I pulled out, the money should be spent directly [on the ground]. (Neville, Downside)

In the Mitchell River group, issues of belonging and identification were construed slightly differently to the other groups, largely because of the physical space separating members across the catchment. Non-involvement was largely a function of distance. Owing to the huge area and the rough nature of the terrain, frequent road travel was expensive and time-consuming, especially in the wet season. Air travel was the preferred method of getting from A to B but, given the high cost, this means of travel was not accessible to all members. Where possible, individuals would endeavour to share the costs of travelling to the same meeting or field trip. Another solution to the problem of isolation was to rotate meeting venues between the eastern and western parts of the catchment.

In addition to belonging and identification, membership is also determined by personal investment. Members of community groups who invest personal time or resources in the operation of their group will value their membership highly and develop a strong commitment to the cause. This investment in turn will have a significant effect on those members' sense of community. These observations hold true for each of the three stewardship groups.

Investment includes contributions in financial terms as well as the amount of time members spend working with the group. On average, members of the Downside Landcare group invested more financially than did members of the other groups. Each Downside Landcare group member paid a $10 annual membership fee (which was largely perceived as a token gesture). In addition, some Executive Committee members contributed substantial sums of money out of their own pockets:

> In the first year, it was *a lot* [of money]. There were phone calls, trips to town, landcare dinners, looking at sights, lunches, Michael's time, we'd pay for extra things then, incorporation fees, trees, stakes, tree-guards, etc. I wouldn't like to say, it could run into a huge amount of money. (Ellen, Downside Landcare group member)

Most members discounted the contribution they made to the group in lost wages, petrol and donations of food. Many Downside Landcare group members felt uncomfortable talking about the amount of money spent on landcare related activities. However, they all said that "of course it had been worth the effort".

Neither of the other two groups charged a membership fee, although individuals in both groups went to some lengths to contribute financially, either in procuring the necessary materials or meeting travel costs. The cost of attending meetings of the Mitchell River group was particularly high. One person estimated that hiring a plane cost $800. Another member spoke of lost income valued at $2500 over the two years of his involvement, and one Water Watchers member contributed significant amounts of personal resources in analysing water quality samples.

Reinforcement

Reinforcement in a community group context refers to fulfilment of both group and individual needs while maintaining a positive feeling of togetherness. Psychologists have identified several reinforcing features of the individual-group relationship. These are the status of being a member, the success of the group, individual competencies within the group and shared values within the group.[7] I found that the status of being a member has less to do with the hierarchy in stewardship groups (ordinary member, committee member, paid-up member and so on) than it has to do with simply being included in the group - more a function of membership than reinforcement. Two newer members of Water Watchers provided an exception to this. They spoke disparagingly about the position of some of the original Water Watchers members in the local community.

> The hoity-toity brigade didn't get their feet wet ... They knew each other all through the Local Conservation District Committee ... [There was] too much name-dropping and not enough sharing of information. They were all community-minded before [joining LCDC]- they represented some things from before. But communications weren't coming back down the line. They knew but we didn't know and we felt powerless. (Janine and Lucy, Water Watchers group members)

This response suggests a lack of reinforcement derived from a feeling of powerlessness and lack of stature, which in turn lessened their sense of community.

Individual member's skills are factors that reinforce a stewardship group ethic. For example someone who has report-writing, lobbying or tree-planting skills and contributes these to the group helps maintain and reinforce a sense of community. Personal attributes such as enthusiasm, energy, motivation and intelligence also contribute to effective individual-group reinforcement. It seems that people are attracted to groups comprised of individuals whose competencies and personal qualities they admire. The diversity of group members' skills and attributes is a desirable characteristic of group dynamics, contributing to reinforcement for all stewardship groups.

Group reinforcement does not necessarily come from shared values, needs and goals. All three stewardship groups I visited formed out of a common concern with an environmental threat. People joined Water Watchers to monitor water quality and fend off policy makers and politicians who were laying blame at the feet of the Serpentine Jarrahdale community. With the Downside Landcare group, farmers were concerned at the threat of spreading dryland salinity. They rallied to prevent further land degradation on their properties. And in the case of the Mitchell River watershed group, individuals came together to start a catchment management group in order to start thinking about the health of the Mitchell

River; an endeavour of true stewardship, but not at all motivated by shared values.

Sense of community may be strengthened by group members holding shared values - but the basis for these shared values differs enormously between geographic communities and communities of common concern. Stewardship groups by definition are all firmly based in locality, whether on a grand catchment scale such as the Mitchell River group, or around a local government boundary, such as Water Watchers. The evidence and experience from these stewardship groups does not support the assertion that shared values automatically leads to greater reinforcement of individual needs and hence the development of a sense of community. Consider the Mitchell River group. Part of the ideology of the far north is that life is different on the Cape. Qualities of individualism, chauvinism and self-reliance are echoed through colonial history to the present day. However, differences in the assumed value base of members of the Mitchell River group could hardly be wider. It is a question of the degree to which stewardship values are shared which contributes to reinforcement of the group, rather than a simple positive correlation. Later in this chapter the hypothetical link between shared values and group action is exposed.

Emotional attachment

There are many similarities between an emotional attachment as it relates to a sense of community, and an emotional attachment to place. A sense of shared history was strongly evident in the Mitchell River and Downside groups, but was not a factor uniting the Water Watchers group. In Downside, emotional attachment to place contributed in important ways to a sense of community. The "district" forms the basis of a strong cultural attachment between the residents, the land and its settlement history. Many members spoke to me about joining the landcare group for the sake of the district. Raymond said: "If you can get the district behind you, you've got everything". The implied solidarity is supportive of the Downside community's identity. Bill and Madge identified emotional attachment as the main reason for wanting to join the landcare group: "because it was a district project, to give community spirit, [and besides which] socially it was good, they're nice people".

The social networks within each of the stewardship groups support the observations about the interdependency of emotional attachment and sense of community. The more people interact with each other, the stronger their emotional attachment to each other and the group. Conversely, the stronger the sense of community within a group, the more likely it is that emotional attachments will form among group members.

The relationships of members of Water Watchers within and outside the group are illustrated in Figure 6.2. Underneath the oval of group membership, nine circles represent members' other interests and memberships of other community groups. There was little overlap between members belonging to one group and those who belong to another. Above the Water Watchers oval, the oblong shapes represent Water Watchers' formal relationships with other organisations. The group was a subcommittee of the Local Conservation District Committee, an organisation of local landholders, almost all men, who were the umbrella group responsible for accounts and submissions. They were linked to the Serpentine Jarrahdale Shire Council which strongly supported the group via provision of in-kind resources and a venue for meetings. The Shire Council also had strong ties with the Community Catchment Centre which provided facilitation to the group. At the top of the figure, the Serpentine Jarrahdale Ratepayers Association was obviously linked to the Shire and was also linked to Water Watchers with several members in common in the two groups.

Social networks for the Downside Landcare group and the Mitchell River group are shown in Figures 6.3 and 6.4 respectively. In Downside, Landcare group members were very likely to also belong to other local groups and there was less discrepancy among the types of these other community groups. The four organisations supporting the landcare group held less formal relationships with the group than the equivalents in Water Watchers case. Consequently, emotional attachments between all group members were more likely to form. Not only was contact between members more frequent, but the interactions were qualitatively different from those of Water Watchers due to the broader focus and narrower membership base of the supporting groups.

Figure 6.2: Water Watchers' community network

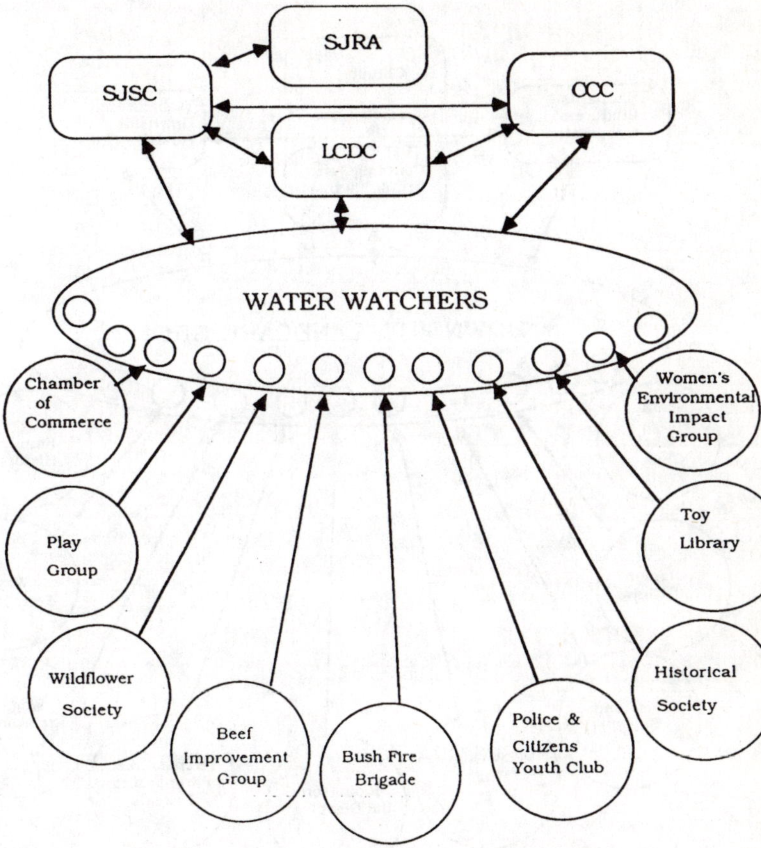

SJ = Serpentine Jarrahdale, SC = Shire Council, RA = Ratepayers Association
CCC = Community Catchment Centre, LCDC = Land Conservation District Committee

Figure 6.3: Downside Landcare group community network

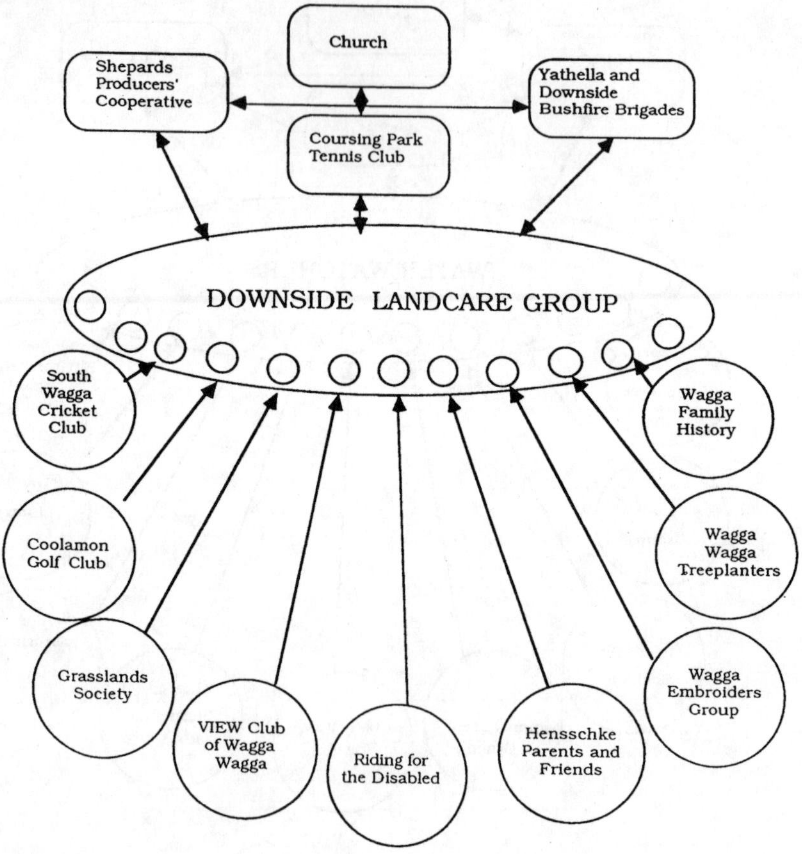

The role of the church in Downside was an important feature in the formation of spiritual bonds leading to emotional attachments among some landcare group members. The history of Downside is strongly linked to the Presbyterian and Lutheran churches. Many members of the Downside Landcare group were closely affiliated with one or both of these churches. The spiritual dimension and networks built within the Downside community on Sunday mornings seemed to reflect some of the stewardship ideals or land ethics brought to landcare group meetings at other times of the week.

Links between the role of the church and landcare within the Downside community strengthened this attachment. Daphne said "being able to bear witness to other people about landcare" was a high point of her involvement. It seems that the strength of community founded in church attendance was reflected in attitudes to the land for some landcare group members. Majid Rahnema could easily have been referring specifically to the Downside Landcare group when he wrote:

> The sense of sharing common spiritual ideals of a purifying nature can create new and contagious forms of enthusiasm and solidarity, which in turn greatly increase the operational effectiveness of the group.[8]

This spiritual cohesion is also captured and expressed by community psychologists as one of the contributing features of emotional attachment that leads to a stronger sense of community. Spirituality and religion for them involve a sense of transcendence, of going beyond daily life and therefore of coping better with external stress.[9]

Daphne who spoke of witnessing about landcare also spoke about the links between her beliefs in caring for the land, caring for children, and her spiritual philosophies. The strength of her conviction and the elegance with which it was expressed is comparable with the writing of some ecofeminist philosophers and deep ecologists. Unwittingly, she paraphrased Chief Seattle in his famous treatise on sustainability, our planet and its people when she said:

> We should nurture our land like our children, we should pass it on in better condition than we received it. We are just stewards really, it's not our land, just like they're not our children, just leased to us.

Emotional attachment within the Mitchell River group was much less evident than in the Downside group, see Figure 6.4. Neither the formal or informal organisations supporting the Mitchell River watershed group had many shared members. Members were socially as well as geographically isolated from each other. For the most part, Aboriginal peoples, miners, fishermen (there were no professional women fishers), graziers, tourism operators and environmentalists have kept to their own camps, rarely mixing socially outside working group meetings.

The central means of maintaining a sense of community in Australia's north is to keep in touch with local news and information through the bush telegraph. The Mitchell River group seemed to help perform this function. As mentioned in Chapter 4, members' motivations for joining the Mitchell River group had much to do with finding out what was going on, having a say and making sure that others did not have too much say. Before telephones were installed, the mail plane distributed not only letters, but also gossip and news from other places in the region. In the same way, joining the Mitchell River group served to fill members' needs for information.

Cohesion in the Mitchell River group was severely constrained. Despite the best intentions of the original group organisers, members were limited in their sociological and geographical abilities to work together when planning events and were therefore also restricted in their capacity to form emotional attachments.

Influence

John Turner describes social influence as follows:

> The key idea in understanding what researchers mean by social influence is the concept of a *social norm*. Influence relates to the processes whereby people agree or disagree about appropriate behaviour, [and the processes whereby people] form, maintain or change social norms.[10]

There are two ways in which social influence is pertinent to the stewardship groups in question. First, the ways in which individuals influence the processes of the group, and secondly, the way in which the group influences individual members. It is important to reiterate the finer distinction between reference groups and membership groups alluded to earlier. Reference groups are those with which individuals compare themselves to assess their own situation, to refine their attitudes or obtain guidance about their behaviour. They may be defined geographically – but are usually comprised of communities of

Figure 6.4: Mitchell River watershed group community network

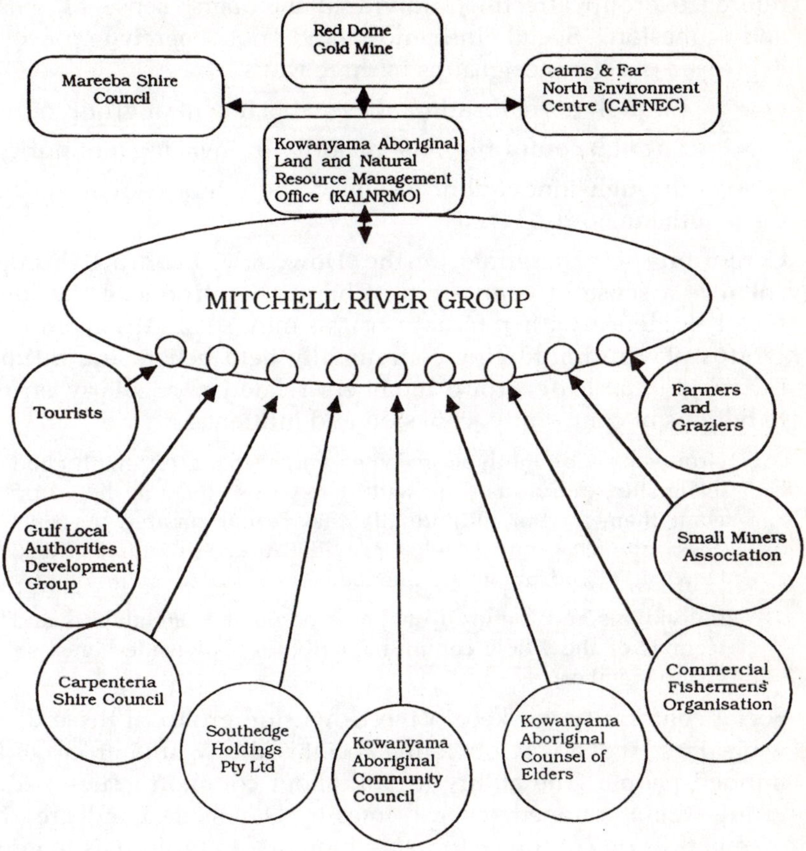

common concern. Membership groups, on the other hand, are those which an individual is in by paying a fee, for example, but which they may not use to compare their beliefs and values. Of the three stewardship groups, the Downside Landcare group is the only true reference group affecting individual members' sense of security and cohesion. Social influence (and not coercive power) is expressed in reference groups in three ways:

- through normalisation, the establishment of group norms;
- through conformity, the majority influencing a minority;
- through innovation, in which the minority may influence the majority.

Conformity is operating in the Downside Landcare group to enhance a sense of community. Chapter 4 introduced the idea of social influence with reference to the minor role that community groups play in reinforcing communally held beliefs and attitudes. Downside Landcare group members related particularly strongly to notions of conformity, cohesion and influence.

> Groups provide motivation. When you're in a group with good leadership, individuals are willing to do as others in the group want them to, but individually they're not capable. As time goes along, groups help provide peer pressure. (Rick, Downside Landcare group member)

> Individuals and the group act as a barometer of thoughts and feelings of the whole community. (Kevin, Downside Landcare group member)

Social conformity worked in the Downside group in three distinct ways. First, it is a way of testing social reality with a group of like-minded people. The ability to talk about common issues within a group was mentioned several times by Downside Landcare group members as one of the factors that motivated individuals to join the group. Secondly, having joined the group, peer pressure toward uniformity of ideas and cohesion of beliefs and behaviour strengthens group conformity and identification. Thirdly, the tendency to check out other members' beliefs and ideas increases when information about the biophysical realities of the community declines or becomes overly complex. This is known as the consensual validation construct which "assumes that people possess an inherent need to know that the things they see, feel and understand are experienced in the same way by others"[11] and which elegantly introduces the topic of local knowledge.

Local knowledge

Knowledge is socially and culturally constructed and grounded within everyday experiences and emotions. Therefore the importance of local knowledge is paramount to the experience of environmental stewardship. This section first investigates how people become aware of environmental problems. Secondly, I briefly review the theory behind ways of knowing, and thirdly, I analyse traditional ecological knowledge as it relates to stewardship groups.

Awareness and recognition of a problem

Environmental problems are not mere conglomerations of biophysical symptoms. They are manifestations of social, economic, cultural, physical and ecological pathologies. "In other words, environmental problems are not so much "out there" in the environment but exist as a function of the relationship between community and environment".[12] In addition, environmental issues are complex, involve a degree of uncertainty and are constructed in different ways by individuals compared to groups (see Figure 6.5).

Reference groups such as the Downside Landcare group construct shared meanings around the multiple realities of living in Downside. However, there is a marked difference between a group mutually agreeing to a set of environmental problems and individuals constructing the same problems based upon their unique values and attitudes. For example, in Downside, widespread dryland salinity poses a serious threat to the future of farming and motivates many people to join the landcare group. However, not all farmers in the district acknowledge salinity and its associated problems as severely erosive forces in terms of productivity, ecology or the social cohesion of the community. Kevin said:

> We weren't really aware of The Problem until the landcare group; we weren't prodded into action. But we've been watching The Problem ever since we've been here; it's only exacerbated into a real worry over the last fifteen years. Some people say it's not a problem and ignore it, [at least] not a worrying problem, but even they are gradually coming around to realise we've got to do something about it.

Figure 6.5: Complexity of key environmental issues in Downside

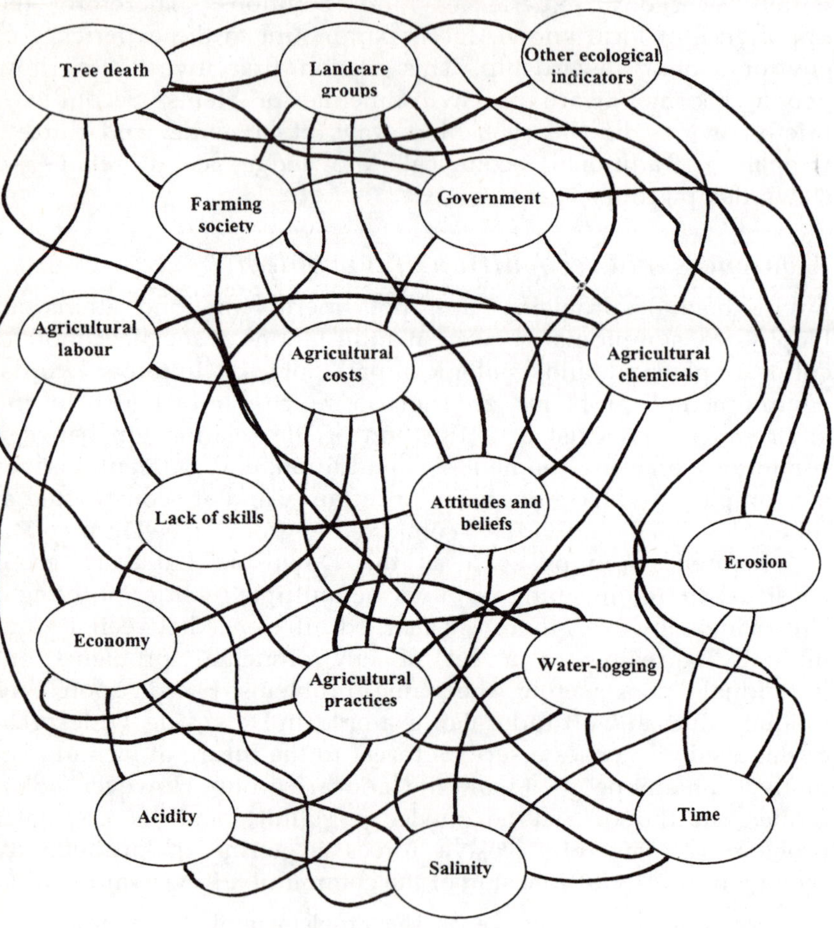

Rising water tables are the problem to which Kevin was referring. There are a number of other land degradation issues of concern to Downside's landholders and to officers of the Department of Agriculture. These include surface waterlogging, soil structure decline, dryland salinity, tree death, pests and weeds, nutrient decline, soil erosion and soil acidity. However, landholders' concern about these problems varies considerably.

Even naming "the problem" creates difficulties for people like Kevin as exemplified by the quote above. Paulo Freire recognised that for some rural people simply naming a problem meant that they were more effectively able to construe it as something over which they had some power to act.[13] There were some landholders in Downside who remained unable to deal with the issue of salinisation to the extent that they were not even able to name it; therefore they frequently referred to "the problem" as an entity unto itself.

After awareness, recognition of land degradation problems is perhaps the next step toward collective stewardship action. There has been much empirical work in rural sociology attempting to link attitudes and responses to land degradation. However, there has been no direct link found between attitudes toward and behaviour concerning environmental degradation. Environmental psychologists writing in the early 1970s suggested that the perception of the environment forms a screen between attitudes and behavioural responses. More recent work recognises that there are cultural, financial and affordance screens which also mediate between attitudes and behaviour.

The field of environmental perception and social construction is complex and does not lend itself to simple prescriptions. However, while attempting to make this point, one researcher exposed his own urban intellectual roots when conducting a survey on conservation agriculture.

> Farmers are not generally aware of their attitudes to the environment. They think that surveys like this are a waste of time and do not conceive of or intellectualise their responses in the way that urban professional people might.[14]

The same author also conducted research into the relationship between awareness, attitudes and behaviour of landholders and environmental degradation problems. He concluded that:

> most farmers believe land degradation to be a serious problem as a general issue, but fail to perceive their own farms to be at

> risk. There is an ignorance of the early warning signs of land degradation, brought on by the presentation of land degradation solely in dramatic forms in the general media.[15]

In several instances in Downside, it appeared that some landholders who did not acknowledge salinity problems on their own property were not unaware of the problem; they were choosing to deny it or rationalise it. One very common rationalisation was that salinity is part of an historical cycle: "The problems here have come and gone before" as Neville said. Therefore, there wasn't anything to worry about: dryland salinity would disappear in the same way that it had appeared.

Increasing environmental awareness was one of the more common stewardship roles perceived by group members. In the Downside Landcare group, John, Ellen, Tim and Murray among many others spoke very highly of the group's success in creating more local awareness about dryland salinity problems. However, there remains a dichotomy between landholders who recognised that their land was affected by salinity and others who asserted that they didn't have "the problem".

Members of the landcare group often expressed anger and resentment toward non-members who refused to join the group or take individual action to prevent further salinity outbreaks on their own properties. In members' eyes, having recognised the existence of "the problem" and having transferred that recognition into their own personal experience, the next logical step is collective action. But this response does not take into account that different people read the world in different ways.

To unite against a common foe is one of the central tenets of collective action. But collective action based upon collective knowledge cannot come about among membership groups or reference groups *unless there is discussion about the nature of the problem which leads to a shared understanding about how it arose and what can be done about it.* Stewardship groups are not usually reference groups, and if they do provide a backdrop against which people compare their own lives and understandings of the world, collective goals are not established merely on the basis of having shared values. The universally acclaimed goal of a clean environment for these groups is more a dream than a reality.

Underlying this complexity in the relationship between environmental attitudes and behaviour for stewardship groups is the idea that these are not simply individual concerns. Common beliefs about the environment are socially constructed within the safety of a group engaged in the development of a "communicative catchment".

> The communicative catchment is one where people become active and caring in their community, environment and catchment. For this to occur, there needs to be identification with an area and with its people; participation in problem identification and decision-making; and ownership of the planning and action. It is one where differing perceptions of people are recognised and incorporated into the community through debate, discussion and discourse.[16]

Ways of knowing

Not all members of stewardship groups join with the expectation of learning new skills, new ideas or new ways of handling problems. However, learning is the most frequently cited personal outcome arising from involvement with a stewardship group, followed closely by the satisfaction and enjoyment of being a member.

> I'm getting a lot [out of the group], understanding how things are, with support from the group and then we'll get more involved when we understand. If we don't understand anything, then it makes it harder for us [to get involved]. (Charlie, Mitchell River group member)

> It has educated me regarding environmental issues, I know a lot more than I did. (Ellen, Downside Landcare group member)

> I've learned new skills through sampling and profiling on the technical side of calculating [phosphorus] loads. (Robyn, Water Watchers group member)

Adult education theory sheds light on the ways in which members of stewardship groups learn. For these learners, group membership offers a way to find solutions to real environmental problems at the time the problem has presented itself, not before they have had exposure to it. In other words, there is a practical focus. In this case the learning is a by-product of group membership. And the learning is predicated upon the experience of group members in a way which is determined by each group member, not a teacher or a lecturer or an agricultural extension officer. The learning is totally controlled by group members in an environment which creates equality, respects diversity, draws upon individual experiences, facilitates shared responsibility, incorporates experimentation, accounts for emotional attachments to places and people within the group, and encourages social interaction. Indeed, some theorists believe that social interaction is the key to unlocking the whole social learning context.

Social learning theory rests upon the notion that individuals' behaviour in small groups is culturally determined by group activities, which mould the individual, and, in turn, the group. Social learning is a collective experience largely determined by the relationship between the individual and the group, and is especially effective in making sense of complex, rapidly changing circumstances. This type of learning is effective because small group environments actively support their members, provide a strong sense of community and provide a safe context for experimentation. This was certainly true for both the Water Watchers and Downside Landcare group.

Social learning relies heavily upon social practice or experiential learning in a group. Also known as action learning, it can take place individually as well as in a group. Socially active learning has the potential to change behaviour and attitudes within a group. It implies a continual process and is often likened to a spiral where observing, abstracting, planning, acting, and reflecting or reviewing take place within a specific experiential context. The benefit of a socially active learning style for stewardship groups is that these stages can be discussed among group members. This has the added potential to lift the level of learning from individual practice to emancipatory action learning in which all group members share responsibility for environmental stewardship. The group process enriches the quality of reflective action such that "improvements in what we do ... can come about by changing the way we make meaning rather than just reinterpreting the meaning we usually make".[17]

Members of the Water Watchers group listed "doing something" as the primary reason for joining the group. "Getting results back and seeing something *concrete* for the year's work" was listed as one of the high points of involvement in the group. Following the principles of action learning (albeit unwittingly), another water quality monitoring group in Albany, Western Australia, adjusted the tools required to monitor stream flow rates. Members of the Water Watchers group heard about the use of this new technology and accordingly adjusted their own measuring techniques. They had used oranges (which float) for the purpose of calculating rates of stream flow. In Albany, gumnuts had been tried and found to be an equally valid instrument, and not nearly as expensive as oranges. They were therefore adopted as a new tool in the Water Watchers' kit bag of monitoring techniques.

In the case of the Downside Landcare group, members gained a lot of agricultural information from each other. They also depended upon the extension services of the Department of Agriculture and the Department of Conservation and Land Management. On the basis of extension information delivered either in person or through newsletters and information sheets, group members were able to trial new ideas. I observed one instance of socially active learning during a bus tour of the area. Bus tours were a regular feature of group activities to disseminate information and monitor local conditions. They also served to reinforce group cohesion.

On this occasion Downside Landcare group members wanted a bus tour to inspect trials and demonstration sites on local properties. At a minimum tillage demonstration, the landholder reported that he had read about the application of nitrogen to a mature wheat crop to increase its capacity to produce protein. He told the group he had decided to trial this on one section of his crop and asked for ideas about how to apply it. After a lot of animated discussion, members agreed that the best way to apply nitrogen to a mature crop was via aerial spraying. Several other group members decided to do the same thing on their own properties and suggested hiring an aerial crop duster. The arrangements were made there and then in the paddock at the demonstration site, with three or four individual landholders agreeing to share the cost of hiring the plane.

On each of my visits to the stewardship groups, I asked members to reflect on what they personally had got out of their stewardship experience. Typical responses included: "learning about different species of salt indicator plants, plant identification" (Mike DL); "It makes you look a bit closer at your land … it makes you take a bit more notice of things and gives you more ideas [about] what you can do" (Rick and Jill DL). Members spoke to me of increased opportunities for exchanging ideas and some implemented stewardship projects on private properties as a result. The landcare group in particular was a prop for social learning.

To summarise:

- Community groups are seen as useful and valuable sources of information and advice by members.
- Non-members of groups do not have the same access to this type of informal extension and support.
- Through discussion and exchange, groups engage in socially active learning in a small group context which is most conducive to their culture and local needs.

Local and traditional knowledge

The role of local knowledge for groups engaging in environmental stewardship is critical. Local knowledge incorporates information about local conditions, local culture, intuitive concern and local places. An aesthetic appreciation of community should also be added to this list on the basis of the three stewardship groups' experiences. Local knowledge is especially important for people coming from outside local communities, ie. external scientists, historians, artists, anthropologists, journalists and many others. Sheep farmers' local knowledge contributed to scientists' responses to the radioactive fallout from the Chernobyl disaster.[18] Australian farmers' knowledge of their own catchments has contributed to deepened understanding of dryland salinity.

Knowledge networks disseminate knowledge across space and time more effectively than hierarchical systems and "a catchment can be thought of as sets of communications and interactions between actors involved in different knowledge networks".[19] The concept of a network or web expresses the complexity of local knowledge production and dissemination. Water Watchers' ultimate success in terms of collective action was jeopardised in part due to a breakdown in communication between members of the network.

Stewardship group members understand where a community is "at" and act accordingly. This implies knowledge of the local landscape and conditions, as well as knowledge of individual members' abilities. Members of Water Watchers felt that there was no point adopting broad environmental management strategies based on elaborate scientific evidence and regional planning jargon if the community group was not in a position to appreciate this type of approach. Unfortunately, hierarchical strategies of this type were used for 15 years in the Peel Harvey region.[20] By early 1992 inappropriate management strategies and non-participative planning mechanisms had weakened any credibility local people may ever have placed in governments. This effectively removed most Commonwealth and many State government agencies from their local knowledge networks.

Local government on the other hand acted in concert with Water Watchers as just another player. The Serpentine Jarrahdale Shire Council used the existing knowledge of the community to develop a skills register of local residents. During a search conference held with a variety of local community groups, people suggested that in order to make better decisions about local environments, it was necessary to have a list of community assets. This did not stop at the level of services, land and capital items, but also comprised residents' skills and expertise.[21] In a group with a relatively weak sense of community, this register of human resources may have assisted Water Watchers to become more self-reliant. However, it did not come in time to prevent the public education campaigns about nutrient enrichment to continue legitimising the government-sponsored hierarchical systems of knowledge dissemination against which the group reacted.

Aboriginal members of the Mitchell River watershed group also emphasised the importance of group networks in preparing young people to look after the catchment for future generations. Toward this end, the Kowanyama State High School and the Kowanyama Aboriginal Land and Natural Resource Management Office developed a special curriculum focusing on the importance of the Mitchell River within the whole catchment. The curriculum was designed and taught in conjunction with the Council of Elders in an effort to recognise their traditional ecological knowledge. It was assessed by the Queensland Department of Education and registered by the Board of Senior Secondary School Studies of Queensland as a junior high school subject. The proposal states:

> This community considers the retention and continuation of traditional knowledge and skills as a means for affirming the Aboriginality of its community. The course seeks to prepare Kowanyama youth to take their places as future natural resource managers and users ... This program recognises:
>
> - the legitimacy of Elders as teachers;
> - the importance of physical geography for the culture and lifestyle of the Kowanyama people;
> - the need to link past and present Aboriginal "care for country" with future management practices;

- the necessity for relevance in school curriculum;
- the importance of stimulating relevant experiences and contexts in language learning.[22]

Charlie of Kowanyama expressed his disappointment at the lack of education and training opportunities for Aboriginal people:

> They don't know what to do, Aboriginal people are not as brainy as white people; but they got to know! Young people must have proper training. In the near future it will be their future they're caring for.

While recognising that European systems of education are also necessary Vic spoke of the importance of recognising indigenous knowledge systems.

> Getting acceptance of each other's expertise [is a problem] when talking about technical and practical knowledge and the gap of expertise between scientists, graziers and Aborigines regarding expert and local knowledge.

Traditional ecological knowledge may be defined as:

> indigenous understandings of "natural" processes (systems of relationships involving plants, animals and various supernatural and environmental factors) ... It is a knowledge that has an immediate relevance and application for specific societies in particular environments.[23]

Traditional ecological knowledge is always evolving. It is ecological in the sense that biophysical knowledge is integrated within cultures. In 1993 in Kowanyama an event took place that emphasised the importance of elders maintaining youth training in ecological knowledge. A traditional ecological practice misapplied in a modern context by younger Aborigines contributed to an ecological disaster. One of the traditional methods for catching fish in the river and waterholes was to "poison" them with a mixture of leaves and root materials from specific native plants growing alongside the river. Fish would be stunned by the effects of the plant chemistry and float to the surface, allowing an easy catch with no less palatable consequences for those eating the fish. One day, several high school students believed they could catch fish from the Magnificent Creek lagoon in the same way. However, instead of using the traditional methods and materials derived from native plants, they used several tins containing dieldrin and chlorpyrifos taken from the agricultural chemicals shed at the council. The whole ecosystem of the lagoon was upset, many fish and shellfish died and people were advised not to eat anything fished from that lagoon. More than a year later there were still traces of chemical toxins in the fatty tissues of resident fish. However, there were no chemicals in migratory fish and populations had returned to normal.

There is a happy end to this story. After the event, members of the Mitchell River group rallied in support of the community. The Kowanyama Aboriginal Land and Natural Resources Management Office implemented ecological monitoring and awareness raising through publicity materials and educational seminars with high school children and local Aboriginal elders. This event can now be recognised not as an act of vandalism, but as one attempt to apply modern technology to traditional fishing. There are many, more positive examples of the use of traditional ecological knowledge by Aboriginal members of the Mitchell River watershed group to educate non-Aboriginal members.

In the Mitchell River watershed, traditional ecological knowledge is evolving in Aboriginal communities, on grazing properties, in mining camps and in commercial fishing operations. Traditional does not necessarily mean indigenous; it could also refer to folk knowledge such as that used by agrarian societies. Traditional ecological knowledge can provide a unique perspective that complements and extends universal scientific knowledge. This point is elaborated in the next chapter.

Outsiders such as anthropologists have been barred at times from recording the traditional ecological knowledge of some Aboriginal communities. "Knowledge is power, often used to justify a policy of secrecy, of withholding information from competitors or other outsiders" according to one anthropologist[24] who also believes that traditional ecological knowledge belonging to indigenous communities should be shared with scientists from external institutions in a co-operative intellectual endeavour. However, the power relations underlying such co-operative efforts require careful negotiation lest scientists be accused of using coercive means to uncover traditional knowledge.

Self-determination

> The empowerment approach ... places the emphasis on autonomy in the decision-making of territorially organised communities, local self-reliance (but not autarchy), direct (participatory) democracy, and experiential social learning. Its

> starting point is the locality, because civil society is most readily mobilised around local issues.[25]

Race and class, decision-making, equity in terms of gender, in-group politics – all of these fit well into any discussion of self-determination within a stewardship group. They are all equally important when applied to relationships outside the group too, especially between community groups and government organisations, because these issues affect the ability of group members to act collectively and are the fundamental bases of democratic society. However, they will not all be discussed in detail. Here I prefer to categorise them into intra-group issues for an analysis of horizontal integration and extra-group issues for an analysis of vertical integration.

Horizontal integration – equity

Horizontal integration refers to the relationships between social units operating at similar scales, for example landcare group to bush group or church group to a local disability association. The type of integration is egalitarian in nature with flat power structures such that power is not concentrated in the hands of a few people. Gender equity within groups is an example of how far horizontal integration has affected rural decision-making structures, traditionally a male domain. Both Water Watchers and the Downside Landcare groups included women.

Because Water Watchers was originally funded through the Office of Women's Interests, early in the development of that group the men involved perceived that somehow they were being excluded. Through the careful and tactful approach of the woman co-ordinating the group, everyone was made to feel welcome. The role of women in environmental stewardship is a central theme for Water Watchers. Janine and Lucy mentioned that they originally became involved with Water Watchers to prove that they were individuals recognised outside the home. Janine said "I want to let people know that I've still got a name". She went on to say:

> It's a personal issue first and then a community issue; women tend to look around for something creative to do to break out of the "bored housewife syndrome" ... [We're] trying something new, branching out of one mould, diversifying our skills.

Other women became involved as they moved from taking primary care of their children at home to a concern with the school and community as their children got older. Some of the women in Water Watchers went on to become the prime movers behind the group. Lucy said that arising from her commitment to working with community groups she became interested in the Ratepayers' Association. Eventually that group became the Women's Environmental Impact Group which shared many members with Water Watchers. Through a chain of associations with the home, the community, the environment and the future, I realised that women in Water Watchers were not separating one issue from another but making a commitment to stewardship in a broad sense: "once there's a willingness to help others and be involved, this gives you a long-sighted approach to what you're living in – the environment".

The women in this group were effectively bridging the gap between socio-economic and biophysical aspects of their environment and working at integrated catchment management on a level far removed from government definitions of that concept. These women felt a great deal of pride in their water quality monitoring achievements. Robyn said:

> The EPA [Environmental Protection Authority] was told to sit up and take notice; they were quite amazed that this information had come from a group of women playing with water in the creeks of Byford.

The number of women Water Watchers and the strength of their motivation to work with the group was also mentioned by government employees and teachers associated with the children's group, Slug Busters. As one school principal said: "My dads I never see, it's always the mums". He suggested that women's involvement with Water Watchers and Slug Busters was generated out of a concern to protect future generations and their environments. Other government employees noted that often women were the ones who came up with the more practical suggestions about solving a specific stewardship problem. This was also true in Downside.

The Downside Landcare group was very progressive in terms of its involvement of women in a traditionally Australian male domain. Affirmative action in attracting women to the committee was unnecessary. There were ten committee members, five of whom were women and five men. Unlike other landcare groups, there was a noticeable involvement of single women and women independent of their spouses. The number of women involved may have been due to the female landcare facilitator, or to the strong leadership provided by the Downside group's secretary and publicity officer, both of whom were women.

Women's involvement in environmental issues is often premised on a vision of society which reflects their concern with the environment (although not at the expense of the economy). This concern is founded on a lifetime, for some women, of working in community services, where social injustice is more obviously linked to poor environments, poor economies and poor health. But whether or not women make better environmental stewards because they are more "nurturing" or "closer to Mother Nature" is a topic that remains outside this immediate discussion. I believe that as a long-held assumption, it remains open to critical inquiry and should not be left unquestioned.

Women in the Downside Landcare group influenced other members to become more involved in their local community organisations. For example, the group developed an important relationship with a local primary school. In rural areas of Australia the inevitable message at the bottom of any invitation to a gathering is "Ladies bring a plate". Downside was no exception, supper was always provided by the women. However, that was a route to strong in-group cohesion since the informal discussion over tea attracted many who were loath to speak in front of the whole meeting and was a big incentive for local people to attend landcare meetings.

When hosting a dinner for a visiting contingent of New Zealand farmers, one previous non-member became involved with the landcare group because she had been personally approached to assist with the cooking. Once she attended the function, she joined the group. She subsequently became an executive committee member of the Downside Landcare group.

While gender-based equality was not the only basis for equity within Water Watchers or the Downside Landcare group, it was certainly the most common. Cross-cultural equity was rarely mentioned and equity across social strata was only alluded to in passing. However, equity in sectoral and geographic representation in both groups does merit closer examination. As discussed earlier, representation was interpreted widely by members of all three stewardship groups. However, it was generally agreed that each group was more or less representative of its region.

Governments consistently question representation in community groups when it appears that one sector or another has control over community resources and politics. Neil Barr and Roger Wilkinson write pointedly of the importance of two-way flows of information between group members and the wider community to increase representativeness:

> Working group members must consult members of the community informally throughout the life of the group, but especially when there is a difficult decision to be made. Only if they make a real effort to keep in touch with community values can working group members maintain this valuable *gatekeeping* role.[26]

Vertical integration – self-reliance and self-determination

This gate-keeping role raises the issue of stewardship groups' relations with other groups and individuals. Vertical integration is the relationship between a community's social units and those outside the community. The focus here is not so much on the *community's* relations with those outside it, but the community *group's* relations with external organisations. Vertical integration occurs to the extent that the group feels able to take control over factors relating to the environmental stewardship of local places. The Mitchell River group is a very good illustration of the extent to which community self-determination was in evidence.

One of the more noticeable factors about the Mitchell River group was members' pride in and attachment to the idea of bottom-up environmental stewardship of the watershed. This is largely due to the group's roots and to the concept of watershed management espoused by conference participants at Kowanyama in 1990. While many of these were government officers, integrated catchment management was not yet government policy in Queensland. Conference participants applauded the move to form a catchment management group for people who lived and worked in the region rather than those from government agencies based in the regional centre of Cairns, which is outside the watershed.

Operating concurrently with the belief in self-determination were the sentiments that "we know what is best for our community, not government" and "we don't want any government people interfering in our affairs, we don't need their handouts". Self-reliance is a large part of the ideology of life on the Cape. Born out of an historic need, it is a strongly ingrained ethic among group members and other Cape York residents. Members believed that their group had been ignored by governments in favour of integrated catchment management committees in more populated areas. This feeling has contributed to a strengthened commitment to go it alone, proving their independence and community ownership. For example, when the Mitchell River group was deciding on a constitution, it could have chosen an off-the-shelf model, the existing Landcare C constitution. However, significant changes were made to that document recognising the group's individuality and the unique nature of the watershed.

The same sense of self-determination was present in the Water Watchers group who were fighting to prove that high phosphate levels in the streams and creeks at the top end of their catchment were not due to local farmers' over-application of superphosphate. It was this self-reliance that injected a sense of "mini-war" (in the words of one landholder) into a group of farmers who started "fiddling with the drains", to the surprise of the Water Authority of Western Australia. The problem was that nutrients were being leached from adjoining paddocks into the drains during summer when the rainfall was at an annual low point. The resulting phosphorus loads contributed to algal blooms in the Peel Harvey Inlet. Having made several unsuccessful requests to the Water Authority to help sort out the problem, local farmers approached the Minister for Agriculture, got his approval and together with the officers from the Community Catchment Centre, sandbagged the drains, thereby producing a very effective dam. Later, the Water Authority worked with the farmers involved and constructed a concrete dam, preventing the phosphorus-enriched drain water from entering the waterways until it could be diluted with the onset of autumn rain.

According to Kevin "the cockies knew it all along". They were the ones who had come up with an innovative, practical and highly effective solution based on the principle of self-reliance. It was this same tendency toward self-reliance in Water Watchers that gave rise to a decision to ask local firms to donate necessary equipment and promote the aims of the group via free advertising in local newspapers. It is well recognised that self-reliance in communities is its own reward. Acquisition of donations whether in terms of goods, services or time often gains a favourable response from official sponsors and government departments. This can position a group well in the eyes of funding organisations should that same stewardship group request funds later on.

Although some members of the Water Watchers group believed they were quite self-reliant, others did not perceive themselves to be powerful determinants of what happened in their community. Those closer to the urban metropolis felt unable to contribute to real change toward environmental sustainability and a sense of futility existed among those who were not members of the group. Jack said:

> We don't want to know about what happens past our front gate. [Our attitude is] "society should do it, it is government's responsibility, council's job, why should I do it?" People don't see themselves as a community or as able to determine their own future.

He was speaking about a lack of self-determination on the basis of his experience with boy scout groups. As a teacher he perceived himself to be isolated in his professional activities. Environmental studies, while a keen interest of his own, was not officially recognised by the school in which he worked. Interestingly, although he experienced futility in his role as scout leader, he spoke highly of the comradeship he felt in another of his community interests, the bushfire brigade. Self-determination is a feature of groups where members are contributing equally to the costs of producing a collective good and not leaving it all on the shoulders of a few.

In the Downside Landcare group, vertical integration was evidently high because members perceived self-reliance not as a result of individual effort, but arising from collective action within the community. Tim, involved with the group through the local university said:

> It's an expression of the sorts of things that people do together, community spirited people working together, not individual and entrepreneurial only. [Landcare is an] example of how to get things done, demonstrates DIY [and it] has helped change attitudes toward the group.

In the past, self-reliance was seen as a desirable if not altogether friendly attribute of rural life. "I'm all right Jack" was the catchcry of farming pioneers who relied on their own resources. Self-reliance and not having to ask anyone else for help is still strongly admired in rural communities. Just as in the early days, when the harshness of the land was reason to knock it down, burn it and plough it over, there is still a tendency for landholders to band together in the face of adversity. However, there seems to have been a subtle shift in rural Australia: from battling the environment and other farmers and having to rely on oneself, to battling the banks or the government and relying on community spirit for support. In Downside, there was a tendency to support and nurture group self-reliance with respect to other institutions along a vertical scale. However group members also demonstrated strong horizontal integration arising from their well-developed sense of community and place.

Summary

A do-it-yourself attitude can reflect a rebuttal of government attempts to undertake measures toward ecological sustainability, a topic discussed in the previous chapter. But perhaps more important than an emphasis on political power, local knowledge, a sense of community or a sense of place is an emphasis on collective action arising out of these principles.

The real crux of collective action is best illustrated by Shirley who said about Water Watchers: "Group action is more effective than individual action in bringing about change". If a stewardship group has shared goals, is action-oriented and the people in the group are energetic and enthusiastic, then these are very attractive reasons to join according to several other government officers speaking about the stewardship groups they were familiar with.

For each group, collective action corresponded to a move away from traditional representative systems of democracy towards one in which group members had more direct influence over community issues. This is occurring through increased numbers of local people joining stewardship groups, different types of people being involved in decision-making and better consultation with governments. The extent to which people are acting together in stewardship groups is an indication of a belief that environmental problems can be solved. Figure 6.6 illustrates a model of the way in which the four principles discussed in this chapter contribute toward collective action by stewardship groups. Perhaps its resemblance to a molecule from organic chemistry

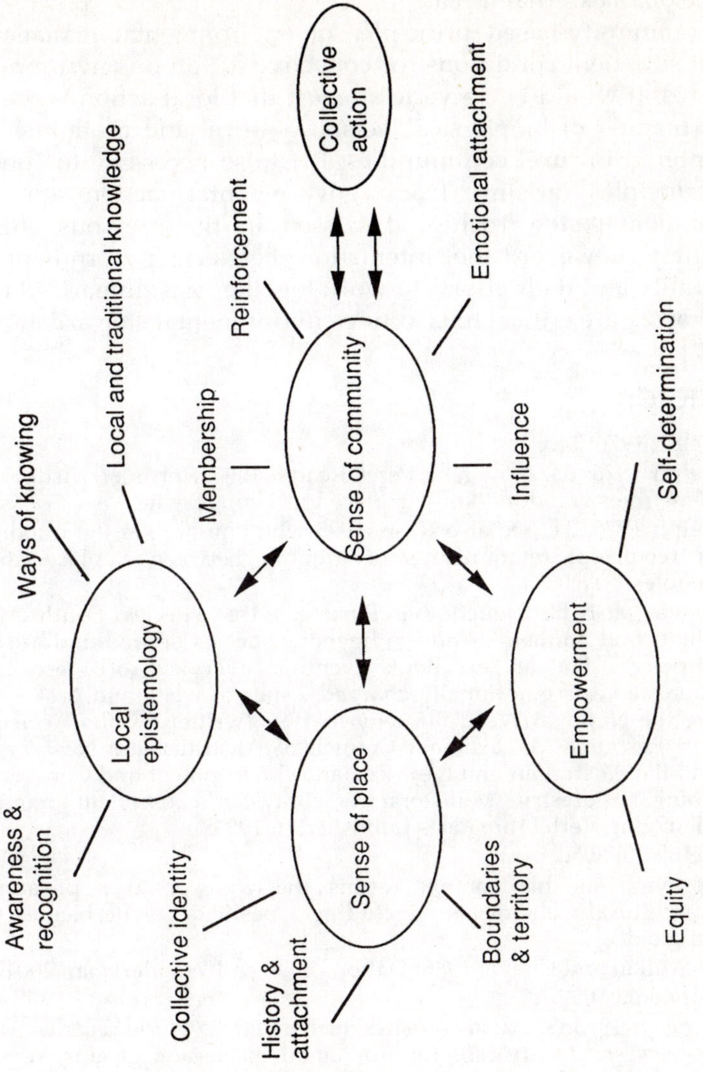

Figure 6.6: Principles for community-based environmental action

reflects community group structure itself, flexible but cohesive in the face of an external threat.

Community-based principles of environmental management are not sufficient conditions for collective action on environmental problems. It would be naive to suggest that local action is the cure for all manner of biophysical, social, political and economic woes encountered in rural communities. It is also necessary to consider the principles arising from governmental action on local environmental stewardship, discussed in the previous chapter. Ultimately however, the interaction between government and community and their efforts to work together as summarised in the next chapter are critical to effective environmental stewardship.

Endnotes

1 Relph 1976: 140.

2 Relph 1976: 33. Note also Peter Read's 1996 words of wisdom, that sometimes you don't know a place until you lose it.

3 Relph 1976: 34. See also Rose, 1995 who emphasises the importance of reciprocal relationships of identity between a place and its peoples.

4 It may also be argued that European tree species, South African plants and animals or non-indigenous species of any kind are "out-of-place". For an excellent scientific overview of these highly contested and emotionally charged issues of weed and pest species, see the National Weed Strategy in 1997 (written by the Agriculture and Resource Management Council of Australia and New Zealand and the Australian and New Zealand Environment and Conservation Council Forestry Ministers) or Braysher's excellent handbook Managing Vertebrate Pests published in 1993.

5 Relph 1976: 50.

6 Spraying in this context means the widespread application of agricultural chemicals including pesticides, herbicides and fungicides.

7 McMillan and Chavis 1986; Dalton, Elias and Wandersman 2001.

8 Rahnema 1992: 171.

9 Sarason 1993. See also websites such as *http: //psychwww.com/psyrelig* or *www.spiritualityhealth.com* for email discussion groups, reference material and information about the relationships between psychology, health and spirituality.

10 Turner 1991: 2.

11 McMillan and Chavis 1986: 11.

12 Martin Pinn, Woodhill and Wilson 1992: 13.

13 Freire 1975.

14 Vanclay 1992: 100.

15 Vanclay and Lawrence, 1995. Other studies support this observation. See Pitt and Yapp 1992; Cary, Wilkinson, Barr and Milne 1993.

16 Martin Pinn, Woodhill and Wilson 1992: 15.

17 Bawden 1991: 88.

18 Wynne 1989; 1992. Wynne has also contributed a lot to discussion about the public understanding of science and the role of local knowledge within this, see Wynne 1995 1996a 1996b.

19 Martin and Lockie 1993: 13.

20 Bradby 1992.

21 Shire of Serpentine Jarrahdale, *Local Rural Strategy Workshop.*

22 Kowanyama Aboriginal Land and Natural Resource Management Office and Kowanyama State High School 1993: 5.

23 Lewis 1993: 8. I chose to ignore the inclusion of folk taxonomies in this definition (meaning the ethnobotanical and ethnozoological classifications of plants and animals) due to the disagreement about what they constitute and how they contribute to traditional ecological knowledge.

24 Hunn 1993: 14.

25 Friedmann 1992: vii. Also, see Dalton, Elias and Wandersman 2001's chapters on citizen participation and empowerment and organising for community and social change.

26 Barr and Wilkinson 1992: 130.

7

Bottom-up meets top-down

Working toward middle-ground environmental stewardship

This concluding chapter analyses both community and government approaches to environmental stewardship and suggests new ways in which dialogue may begin between the two. Therefore, it is timely here to summarise the "cons" as well as the "pros" of each approach, see Figure 7.1. For if my argument is to be convincing – that neither top-down nor bottom-up approaches to environmental stewardship are sufficient for sustainability – the problems of each approach must be fully exposed and analysed.

Problems with top-down approaches

Environmental problems cross local, regional and national boundaries, and so different types and scales of government intervention are warranted. Undisputedly, national governments are somewhat better at implementing certain civic responsibilities and policies than State, Provincial or local governments and vice versa. That government has a role in local environmental management is not in question here. However, I argue that their methods and practices are limited when it comes to facilitating community-based environmental stewardship. Government involvement in local environmental stewardship can be weakened by the rigidity, size and culture of older style "command and control" style bureaucracies, as evidenced by the following problems arising within a landcare context.

Positivism

Some government officers believe that the present environmental crisis is all about focusing on solutions to specific problems – the soil, the water, or desertification, to name a few. Soil acidity problems, nutrient depletion (with concurrent enrichment of the

Figure 7.1: Comparisons between the top-down and bottom-up approaches to environmental management

The top-down approach	**The bottom-up approach**
Potential for: • Awareness of national and international forces shaping local practice • Efficient utilisation and equity in distribution of national resources • Coherence between local, catchment-based plans, administrative assistance • Access to specific technical and research-based information	*Potential for:* • Locally-based catchment planning • Awareness and monitoring of local conditions • Effective utilisation and equity in distribution of local resources • Local action based on ownership of problems • Networking across like-minded groups and people within groups
Middle-ground *The integrating links:* • Effective facilitators • Involvement of schools and other local organisations • Community resource centres • Networked groups and facilitators across regional and State boundaries	
Danger of: • Lack of awareness of local needs and conditions • Difficulty in identifying and coordinating local contributions to national programs • Emphasis placed on large groups, more visible groups and large-scale projects • Departmental and disciplinary-based barriers to effective communication • Short-term politically expedient actions • Simplistic reductionist framing of environmental problems in purely biophysical terms • Institutional and ideological barriers to local participation • Formula/prescriptive approach to community groups	*Danger of:* • Duplication of effort, waste of local resources • Parochial attitudes, not seeing the broader picture • Inappropriate local expectations of achievements • Entrenched leadership not successfully helping the group to progress • Lack of group-process skills and an inability to evolve and mature • Perception that land users lack skills and education required for environmental management

waterways) and tree death for example, can all be fixed by science and technological solutions, according to a minority. One farm management consultant I spoke to expressed the opinion that: "all we need is some new technological innovation or farm management practice and we can go on farming without these landcare groups". This is not so in my opinion. The naive belief that there is some technological fix, which will decrease rural society's reliance on people working together on the land, is profoundly disturbing. Landcare groups are sources of innovative ideas about adapting or implementing farming technologies. There are a number of other difficulties associated with single minded, positivist, narrowly defined scientific solutions discussed below. However, an important caveat to this analysis is that it would be unwise to over-generalise and damage the significant reputations of scientists who have made extensive contributions to specific biophysical problems or agricultural extension officers who embrace post-positivist views of science and the environment. In this instance, I have deliberately chosen to take a community-based perspective on studies of science and the environment. Yet, it does not mean that I wholeheartedly and uncritically support populist anti-science opinion. Simply because I engage in critical examination of the social roles for scientists in science does not mean that I am anti-science.[1]

First, scientific and technological solutions based solely on rational science are not solutions. At best, we global citizens can hope to alleviate the effects of soil structure decline for example, but such complex and widespread environmental degradation problems can never be "fixed". Or, as Daphne from Downside put it: "we can't put it back to better than nature". And, if there is some temporary respite found to problems such as soil acidity, for example by applying lime, scientists still do not know for certain whether it is just a temporary measure – because there is insufficient baseline ecological knowledge to predict what effect current remedial actions will have in the future. In the past, traditional curricula have taught agricultural extension officers that science is certain, predictable and secure, offering a range of rational "solutions" to any one problem. Some top-down approaches to agriculture have seen "solutions" implemented nation wide. For example, years of superphosphate applications on crops and use of leguminous pastures (especially subterranean clover) brought benefits to many Australian farms for decades; both financially, and also environmentally in cases where such pasture improvement was implemented to help control soil erosion. A steady decline in soil pH in the wheat and sheep belts of Australia did not emerge as an acidification problem until recently when it began to impair productivity. Presently and in future, more attention will have to be paid to the risks, uncertainties, complexities and fuzziness of science. Although positivism has gone out of vogue in science practice in the 21st century, its skeleton remains intact. In an era of uncertainty and rapid social and technological change, society tends to cling to ideas of methodological surety, standardisation and proof.

Secondly, the traditional agricultural research culture has tended to strengthen the divide between "experts" and farmers. Farmers, like other "target populations", have seldom been encouraged by experts to come up with their own new ideas and approaches to old problems, despite the fact that they always have and always will. In Australia during the 1950s and 1960s the best extension officers relied very much on a transfer of knowledge from successful farmers' innovations to the wider community. More recently farmers in most first-world nations have lived with the top-down chorus of "ask an expert", or "check it with the Department" first. On the southern slopes of New South Wales in Australia, some landholders perceive that Agricultural scientists still rely on statistical trials and plots isolated from real farms in real conditions, which farmers believe only serves to reinforce the divide between agricultural research and agricultural practice. This practice of "white-peg" agronomy often fails in terms of agricultural extension because a typical statistically designed field experiment provides a very poor basis for extrapolation of results from the plot to other locations and in other seasons. This long standing, but unwritten principle of intellectual distinction has contributed to "dangers that the state will suffocate local initiative and responsibility or capture and harness local initiatives and resources for other purposes".[2] It effectively removes local disposition toward self-determination.

Thirdly, there is mounting evidence of public mistrust in Science. It is documented in Wynne's Cumbrian sheep-farmers, Jasanoff's critical exposé of the BSE ("mad cow") scare in Britain and in many other studies of science.[3] Many of these authors also point out that if the State is not to lose face with its political constituency, it must pay more than lip service to re-establishing public trust in science. Thus research councils and R&D corporations are requiring more effective science communication and more outcome-oriented or practicable products of publicly funded research. These efforts, however well intentioned they may be, are unlikely to combat the significant shift in public attitudes against science and technology. The State must also relinquish some power over science and recognise the substantial efforts in community research, community science and collaborative science policy fora such as Danish Consensus Conferences as extended peer communities.[4]

The fourth symptom of positivism associated with the top-down approach is its fixation with biophysical problems, rather than looking more holistically at the range of problems communities face which encompass social, economic and cultural dimensions in addition to the biophysical ones. Government officers in various bureaucracies have helped propagate the belief that all we need to do is to find indicators of environmental quality at the top and bottom of a catchment, monitor them and implement appropriate remedial steps. Recent American research confirms this – "most public land managers – as well as many of the constituents they serve – have educational backgrounds that are strong on understanding the land, but weak on understanding people".[5] They forget that the environment is not simply comprised of a set of discrete biophysical issues. While tree death, dryland salinity and surface waterlogging are severe problems, all have a socio-economic context and all issues are intricately interconnected. Just as important to the future environmental health of global farmlands are issues of declining rural populations, teenage suicides and lack of adequate education and health services. Community-based environmental stewardship is concerned with monitoring of biophysical entities and protection and restoration of ecology, but it is also concerned with people and their livelihoods which cannot be treated simply as scientific obstacles to be hurdled.

Disciplinary chasms

In recent attempts to work with local communities and groups, extension officers who previously found that their training in agriculture at university or college was adequate, are now desperately seeking systems. Systems, that is, to better understand the interaction between human, cultural socio-economic organisations and non-human, ecological biophysical realities. However, there are a number of difficulties, which prevent interdisciplinary studies from informing local debates about environmental stewardship. For example, universities are typical examples of hierarchical and slow moving organisations. University education is seldom progressive when it comes to preparing students for complex interactions between the land and its people, or to face a bewildering array of environmental issues, personalities, values and timelines. One of the most difficult problems confronting policy makers and university lecturers alike, is how to prepare potential extension officers, facilitators and other students for liaison between government and community groups, especially given Campbell's observation that group facilitators may be drawn from "journalists, archaeologists, teachers, engineers, foresters, horticulturists, small business people, farmers and agricultural scientists". He goes on to state that although almost half of these facilitators and coordinators are women, "most of the state agency staff they are working with are male agricultural science graduates".[6]

As noted in Chapter 4, a committee of landholders from the Downside Landcare group appointed Cathy, a local farmer with an agricultural science degree who was also trained as a teacher, rather than choose someone from the many young extension specialists who applied for the position. The essential difference was that Cathy was a local person with locally developed communication skills and a broad appreciation of the complexities underlying the landcare group, while the young extension specialists, in general, were not. The point here is not that a science background makes you less qualified for the job – it does not. However, science by itself, is not enough to cross some of the disciplinary chasms.[7]

Institutional barriers

To some of the more established public servants in the environment arena, the idea of collaborating or cooperating with a rival department is foreign at best or abhorrent at worst. Some State government departments (not to mention Federal and local government departments) are so uncooperative that stewardship groups suffer from a lack of access to information, fragmentation of decision-making or outright confusion over available resources, to name but a few complications.

A related issue is the extent to which bureaucrats go unpaid and unacknowledged for liaison with other departmental colleagues. I am not suggesting that liaison skills are completely unrewarded. Rather, I am suggesting that despite the best intentions of inter-departmental committees and coordination efforts, there are still differences between departments, (and between policy and administrative units of the same department) which hit at the more fundamental cultural alignments of groups of employees. Indeed, inter-departmental cooperation often makes their job harder and is often not encouraged by government senior executives since: "an enthusiasm for an inter-sectoral approach can jeopardise a career, and within most governments, there is little impetus to liaise with departments competing for influence and limited resources".[8] This is beginning to change with many government departments establishing ostensibly open and accessible policies on the availability of information.[9] For example, consider the legislation in one Australian State stipulating that information *must* flow freely between government authorities. That occurred as a result of a change in political ideology when the long established conservative government was exchanged for a more liberal one. However, there is still more evidence to the contrary illustrating that with the current global moves toward increasing privatisation of government, there is an emerging ethos of competition for information in an era of "new public management". Institutional barriers are alive and well in Australia. Consider the following as a timely reminder:

> Institutions are devices for achieving purposes, not just for achieving agreement ... institutions also contribute to decisions not being taken and to preventing things from happening ...[10]

For example, it is generally established that rural industries face severe constraints when working with centralised and rigid bureaucracies typical of current agricultural institutions. However, it is only in the last decade that research has become increasingly critical of specific inflexibilities in the way those agricultural institutions are run at the local or regional level.[11] The so-called "culture of the old guard" has prevented many new paradigm, participatory approaches to land management from being realised. In Australia, the Netherlands and the United States there have been well-documented attempts to teach holistic, systems-based and people-centred approaches to agriculture and the environment in some universities. However, there are many attitudinal and ideological barriers created by the old guard within other universities and agriculture departments to prevent such change from occurring. Again in the United States, barriers to collaboration "include rules of public participation that have been set by judicial precedent rather than common sense; [and] laws intended to prevent backroom collusion that have made it hard for agencies to seek advice".[12] But, to be fair, it is not only career-minded senior bureaucrats who discriminate against these changes, it is also an increasing lack of resources, among other constraints.

Funding and the messenger state

While environmental non-government organisations in the United States and Canada have a history of independence from government and closer relationships with philanthropic organisations, grass-roots groups in Australia and New Zealand have almost exclusively relied on financial support from government. Of all the problems associated with government policies designed to encourage the publicly funded stewardship groups, perhaps the one which is most widely recognised, is a lack of resources. This is true of projects in their initial submission stage – there are invariably more submissions than there is cash to go around. But it is even truer of ongoing projects. In the past, governments have displayed a tendency to overcommit funds in the first year or two of a new program such that by year three and four there is barely enough to support existing projects, let alone to establish new ones. In Australia, the National Landcare Program was an obvious exception to this. Funding was initially announced for a ten year period (The Decade of Landcare) and was bilaterally supported in Parliament before the program was announced, making it the first such initiative in Australian political history.[13]

Increasing foreign debt, rising unemployment and variable interest rates all form part of government rationalisations for the decision to cut expenditure of new programs in the face of increasing international economic pressure. Surely national government cannot be blamed for their position as messenger bringing the bad news of international agricultural decline down to the local level? Or can they? It is true that some stewardship activities have arisen as a direct result of governments passing their environmental responsibilities onto the community. Some would argue that the problem is not that resources are scarce, but that "dry" economists govern the disbursal of finance and resource security is all in the eye of the beholder. Whatever the political interpretation of this act, it is clear that environmental community organisations are feeling the pinch.

The most disastrous effects of this under resourcing often fall on the shoulders of a group facilitator. Where funding is tight, in many instances the facilitator ends up doing too much work with too little support in too short a time frame. The net effect of this is that often the group's facilitator leaves, and in a worst case scenario, the group itself may fold. Overload and burnout will continue to be a problem among both government paid facilitators and voluntary leaders of stewardship groups unless there is a new infusion of resources to enable people in those positions to do their job more thoroughly, ie with more time and attention to detail.

Change and uncertainty

Partly as a result of a diminishing resource base or perhaps in spite of it, new public management rhetoric has effected a noticeable rise in the number of government departmental restructurings. This not only angers and disturbs public servants, many of whom have just become accustomed to the last readjustment, but it also raises issues of identification, ideology, security and (more importantly) creates dilemmas for the public. An example of this type of problem for stewardship groups operating between 1990 and 2000 is the confusion arising over names. In New South Wales, the Soil Conservation Service, a statutory authority known as Soil Con, was merged in 1991 with the Department of Lands and Surveys, to become the Department of Conservation and Land Management, (CaLM). For a few years afterward it appeared that CaLM did not enjoy the same recognition as Soil Con did and so community members oscillated between names. The confusion was exacerbated when former Soil Con officers continue their personal alignment with the old Service at the expense of the newer department. In 1995 a further merger was announced between CaLM, the Department of Water Resources and the Water Services Policy Division of the former NSW Public Works. The new department was known as the Department of Land and Water Conservation. Yet another example is that of the Victorian Department of Natural Resources and Environment who used to be the Department of Conservation and Natural Resources (CNR) who were known before that as the Department of Conservation and Environment (DCE) and before that as the Department of Conservation, Forests and Lands (CFL). It is not surprising that stewardship group members were confused[14]! Groups such as Downside Landcare, typically operate over longer time periods with relatively few structural changes in comparison to government.

Although this type of change and uncertainty is not inherently "bad", it can be confusing and time-consuming for people at the community level who have a huge range of more pressing uncertainties to face on a regular basis whilst attempting to make a living off the land. It also takes government institutions a long time to establish trust and cooperation with communities and may leave government officers with a sense of unease or discomfort. Importantly, from the government's perspective, a common rationale for these mergers is to integrate resource and environmental concerns in one institution, rather than have them separate. But, while the arguments for such mergers may be solid, in practice they often leave members of the community and staff wondering what has happened or feeling threatened by the impersonality of a large bureaucratic system.

Large-scale (green-field) planning

An international preoccupation with large-scale development in the 1950s through to the 1970s served to marginalise many small scale and locally based projects. Dams and electricity supply projects typified the top-down preoccupation with greenfield environmental planning of this era. In many first world nations these large scale projects provided a problematic signal for community groups. Some "first up, best dressed" stewardship groups still believe that to get in first and set up big is necessary for continued financial security from government. Members of one landcare group I visited had written a series of large scale funding submissions and lobbied local, State and Federal politicians at the expense of conducting local stewardship activities. When the submissions were rejected, not only did group members lose face and momentum locally, they became more and more prone to cooption into State and Federal government agendas. In their bid to become advocates of community-based landuse planning, they forfeited their focus on local needs and opportunities and effectively became a political advocacy group, rather than a landcare stewardship group.

There are three serious concerns regarding government cooption of community groups: 1) financial and ideological dependence upon government for fear of "biting the hand that feeds them", 2) subtle alteration of groups' goals and mode of operation through increased contact with government, and 3) overt encouragement of group facilitators to fit in with government in the hope of getting further employment.[15] This last type of cooption is becoming more common where groups' own needs and concerns (for example with social and educational objectives), become secondary to those of State government departments. In cases where a government department has a legal mandate to coordinate the management of rural resources, for example, that department is not likely to support or finance community-based education programs about the resource since an educative function is perceived as primarily located within an education department. Demarcations of this nature are exacerbated if political itineraries become enmeshed in a group's success. Extreme cooption is evidenced when a community group begins to take on characteristics of the bureaucracy with whom it is working to the detriment of local residents.

These words of warning were issued well before many of the current participatory processes of government were instituted:

> The continuing innocent reliance upon the expertise, anonymity and impartiality of government bureaucracies has not exhibited or encouraged the type of democratic involvement in environmental management wherein our hope for a better type of survival surely lies.[16]

This is not to lay blame or point fingers at government officers who are, after all, only doing their job. Without the kind of support offered by local bureaucrats for local environmental management, it is unlikely that the Water Watchers, Downside Landcare or Mitchell River stewardship groups would have progressed in the way they have toward their specific environmental stewardship goals. Community-based environmental groups are themselves open to criticism and deserve closer scrutiny.

Bottom-up approaches

Stewardship groups have grown in number and are diverse in strategy, as noted earlier. They often have a history of alternative development evidenced by increased local economic capacity, empowerment and other self-organising principles:

> Such local empowerment is the common political thread that runs through numerous contemporary political movements,

> from wilderness preservation, to the future of the family farm, to native self-determination, to the Green movement, and to liberation theology.[17]

Common to them all are barriers and problems arising from their localism. Several of these constraints closely resemble those of the top-down, especially in terms of global limitations due to macroeconomic conditions and international affairs. However, I do not agree with centralist top-down positions that denigrate all local attempts at self-determination. It is too easy to pooh-pooh the attempts of local stewardship initiatives simply because they are caught up in global integration as transnational companies glide across nation states pillaging and plundering local resources. I believe local efforts should be recognised and supported as proactive, practical steps in the face of environmental adversity. The bottom line is that we as global citizens are all responsible for environmental stewardship and should not rest on our laurels simply because it is difficult to rebut the global political economy.

Resource efficiency

One of the frequent criticisms of local approaches, whether they are planning, health or environment related, is their tendency to duplicate each other. This is not an insurmountable problem on its own. However, the accompanying duplication of resource use does create difficulties. For example, water quality monitoring groups in schools often require the services of a government adviser. Unless these groups are coordinated and have the capacity to share personnel, there is a danger that the human resources required to coordinate information will be distributed inefficiently (and sometimes inequitably) among a dozen different schools in the same area.

Another more persistent problem evident in community groups occurs when many groups request Federal Government money for very similar functions albeit in different areas. The result of this multiplication of effort is that many submissions don't get past stage one. Duplication could be avoided here, with more attention paid not only to the unique geographic position of the community but also to the nature of their program of works, via a coordination base in each region. Australian efforts to filter stewardship group submissions and applications have taken the form of catchment management committees and regional assessment panels. However, while integration and coordination are their stated objectives, in practice they too have been criticised for introducing another level of bureaucracy, power and politics into an already unwieldy administrative system.[18] Perhaps multiplication of stewardship projects is not in itself a problem, since they do operate locally and they work on different environmental issues operating at different scales and geographic locations.

Exclusion and belonging

Assuming that equity and diversity are desirable goals, a rather more serious problem for emerging community groups is their tendency to exclude entire sections of the local population. Largely unwittingly (although not exclusively), indigenous people, women, people from non-English speaking backgrounds and in some cases, poor people have been overlooked in the drive to set up stewardship programs. There are some famous cases of rural elitism in country towns whereby group membership for the disempowered and for women is actively discouraged.[19] Since much stewardship group activity is based around social events, it is crucial for emerging groups to be aware of these sociologically delicate issues of representation, gender, race and power.

Acting too locally, not seeing globally

Some stewardship groups suffer from not seeing the global picture and its incumbent effects on the local community. Acting only with local interests in mind leads some groups to band together and act at the expense of other local groups. A typical example of short-sightedness in a bottom-up approach is that of a salt-action group whose members decided to put in a tiled drain to more effectively drain salty water from their sub-catchment into a river where it caused problems for others downstream. "Out of sight – out of mind" operates in a significant portion of local cases where groups may not recognise or fully appreciate the implications of such a decision.

Misunderstanding or ignorance of global issues is not as much a problem by itself as is the attendant waste of local resources if local goals are not formulated in the context of international, national and regional frameworks. "The main dangers of the bottom-up approach arise when local initiatives are formulated in a vacuum, without due consideration of the broader context (eg Provincial, national) in which the community functions".[20]

Of course, stewardship group goals are dependent upon the nature of "the problem" and upon the spatial and temporal context in which it occurs. In terms of climate change and air quality for example, one would not expect many local groups to introduce concerted action on such cross-boundary issues. But such groups do exist (see for example the work of Severe Thunderstorm Spotters or Smogbusters), and they rarely tackle these big issues without consideration of broader institutional and policy frameworks.

Great expectations

The retention of unduly high or unrealistic expectations about what can be achieved given a specific timeframe and set of resources is prevalent among stewardship groups. Perhaps this situation is preferable to one in which they have little or no expectations of what can be done, however, neither situation is optimal. Unreasonable expectations regarding government support is widespread in Australia and New Zealand where there is a long history of reliance on government concerning the environment. When stewardship groups lobby government for resources and opportunities at the expense of local stewardship action, undue strain and pressure is placed on the community. If this occurs, the group chairperson or facilitator may "burnout", and/or the group may become increasingly politicised and frustrated with the resultant inaction and constant disappointment.

In instances where a group has little or no expectation of success, questions group members should ask themselves include: why did this group form and who initiated its formation? If groups are driven largely by one or two people as the facilitator or coordinator, they may not be accurately expressing a widespread community stewardship issue. Alternatively, they may not be effectively canvassing community support. Workshops and seminars on burnout have been conducted to combat problems of overload for facilitators and voluntary group leaders. However, for groups who are lacking in enthusiasm and commitment, careful thought is required on questions like "whose problem is it?", "how well can the group function under these conditions?" and "what does the group expect to achieve?" If these questions are not framed and addressed within a local context, there is little chance of successful local stewardship.

Misery

At a series of meetings with farmers in the southern tablelands of New South Wales, Ted spoke to me of the "misery in the bush" to account for some of the feelings of futility he was experiencing. He was not simply describing farmers reactions to droughts or floods. In addition to the ecological problems of the day, he felt that many complex economic and social factors, were causing an element of defeatism to be more frequently expressed in country pubs and social gatherings. Lack of social cohesion and unwillingness on behalf of some landholders to believe in group process may lead to stress in farming families. With press coverage of issues such as rural youth suicide and the impact of structural adjustment on the family farm, comes an awareness of and an admission that life on the farm is not always rosy.[21] Historically, farming has been constructed as an expression of frontierism especially for those allocated small parcels of land often considered marginal to the purposes of agricultural production under the *Returned Soldiers Settlement Act* 1924. This is clearly demonstrated by this excerpt from historical mortgage records given to me by Mike of the Downside Landcare group:

> The Mortgagor will from time to time take all necessary steps for keeping down and destroying all rabbits and other animal and insect pests and for exterminating all prickly pear and other noxious weeds and all herbage pests from off the said land and for cutting and keeping down all undergrowths, suckers and scrub that might interfere with the enjoyment of the said land.[22]

"Interfering with the enjoyment of the land" sounds like a crime prosecuted by many Australian country women who describe their husbands as married to the land. Nowadays landholders are increasingly experiencing farming crises induced by suicides, divorce, unemployment and the vagaries of the medium price-taking agricultural economy. If it were simply a matter of weed and pest control, one could imagine many rural landholders experiencing unlimited enjoyment of their properties. As it stands, I argue that the landholder and the landscape are both unwilling, but codependent allies fighting an international losing battle with transnational corporations and agribusinesses, but battle they must. In this situation, one can understand the hesitation some farmers feel about joining community groups, exposing family crises and putting effort into forming new social relations which may or may not offer any solace from their existing social, financial and ecological problems.

Size of the group

One of the difficulties for new groups is a rapidly expanding membership base without due consideration of the quality of interaction among members. Without a subcommittee structure or focus for face-to-face work, overly large numbers of people in groups can threaten the stability and long-term continuity of local environmental stewardship action. Obviously the size of a stewardship group affects how communication is conducted, and the satisfaction of members. If a group is comprised of twenty or more, people will have difficulty talking to each other. Domineering social hierarchies may also compromise group effectiveness. Management consultants have long espoused that an optimal number for decision-making and action in groups is nominally between 5-10 people.[23] This can be achieved in stewardship groups through subcommittee structures. Using subcommittees provides a means of involving more people in less threatening contexts thereby giving all members some tasks and responsibility.

In terms of provision of a common environmental good such as reduced salinity or more trees planted across a catchment, "small groups will benefit from their common interests more than large groups",[24] because of the prevalence in larger groups of "freeriders". This term describes those group members who may benefit from the provision of a common good without ever having to contribute toward the cost of attaining it. It is also true for non group members where their personal gain in terms of reduced salinity or the multiple benefits of tree-planting occurs through the concerted actions of a group working toward common stewardship goals.

Successful succession

As stewardship groups evolve, handing over the leadership can become a critical issue. For an effective transition to stability and maturity, many groups need to pay more attention to building voluntary leadership capacity. If stewardship groups are to survive there must be smooth transitions in leadership, allowing a new leader to come forward easily and ushering out the old leader gracefully. In a survey of landcare groups in North Eastern

Victoria, "a majority of 1998 respondents indicated that there was inadequate leadership and management skills training for group leaders".[25] Certainly many Australian rural communities fear the "tall-poppy-syndrome". The story goes that in rural society, the tall-poppy or person who stands out in a crowd is the first to get cut down to size through the unwanted and adverse attention of family and neighbours. Some group members may perceive that they are incapable or ill equipped to follow in the footsteps of an original group leader for fear of recrimination from their peers or respect for an invisible but very powerful rural hierarchy. Others fear the challenge and responsibility of taking over a new role on top of their existing workload and other commitments.

Leadership succession was certainly an issue for the Downside Landcare group. Although the original chairman, Mike, encouraged potential leaders to come forward (indeed he actively campaigned for a new chairperson), for many group members he was so esteemed as a district leader as to unwittingly scare off perceived "leadership challengers".

Lack of skills and education

With participatory approaches to environmental stewardship increasing, so there is more recognition of the value of experiential learning for group members. However, many stewardship group members continue to believe that they do not have sufficient skills or they lack the "right" education in order for the group to function effectively, despite evidence to the contrary. Any first-hand witness will observe a great variety of technical, financial, planning and manual land management skills executed daily within stewardship groups. Yet many rural people continue to believe they need experts to teach them the basics of environmental management. This message, reinforced over decades of Australian agricultural extension, has done untold damage to community approaches to problem solving. But it hasn't entirely dampened the experimental spirit of the landholder. On the contrary, in the agricultural sector there is a great deal of pride at local and regional field days to display local innovation and adaptation of machinery.

Some people prefer to keep their knowledge to themselves and not to encourage the type of cooperation typical of participatory practice. This disinclination to share knowledge actively works against group approaches to environmental stewardship. This is understandable in the agricultural context where farming is not just a way of life, but a business venture, which is increasingly reliant on good quality information for sustainable management. Much has been written about rural people's knowledge[26] but it is questionable as to whether there will ever be enough known about this subject to devise strategies which overcome the ethic of competition enshrined within capitalist nations. Some Australian landcare groups have even raised the issue of competition, favouritism and lack of communication among members of different sub-catchments of the one watershed.[27]

These problems touch only a few of the difficulties faced by stewardship groups. Among the complex array of principles such as sense of place, sense of community, empowerment and equity issues and local epistemology discussed in Chapter 6, there are many barriers to collective action. Until group process is better understood and more efficiently facilitated, it is possible that locally based environmental groups have been wasting valuable resources on inappropriate stewardship tools. Blind faith in environmental restoration and rehabilitation is naive. Awareness of the limitations inherent in populist methodologies embraced by stewardship groups is the first step toward their resolution. Recognition of the diversity of approaches along the continuum of top-down to bottom-up environmental stewardship is crucial in terms of local decisions based upon knowledge of local conditions.

Continuum of approaches

The growth in numbers of stewardship groups does not necessarily indicate a concurrent expansion in awareness of environmental degradation by local communities. However, it is indicative of the success of publicity campaigns such as that developed by Landcare Australia Ltd, a limited liability company designed to raise funding for and awareness of land degradation. Landcare groups are typical of other stewardship groups in that they have arisen out of an expressed desire to combat environmental degradation. They may be atypical of other stewardship groups in that they overtly express the tension between production and conservation goals within their constituency. Conservation goals are the sole province of most other stewardship groups with production goals remaining sidelined as the property of commerce and industry. Despite this, in the discussion that follows, I have chosen to examine landcare groups because they are largely successful and often cited as excellent examples of Australian stewardship.

Landcare groups' efforts toward sustainable development can be interpreted very differently, depending on whose perspective is being taken into consideration, and along what continuum the groups are being assessed. They may be reflecting community experience and/or mirroring government policy; they can be environmental crusaders and/or advocates of increased productivity; see Figure 7.2 for an illustration of these dichotomies. Note, however, that these dichotomous pairs do not imply that they are polar opposites. It does not follow that if you increase production, conservation must diminish or vice versa. Nor is it true that if you have more community action, you have less government policy or vice versa. The central idea here is that there is a plethora of approaches to stewardship. Lying along a continuum suggests that they blend into each other and not that there are any hard and fast lines which separate one group's stewardship approach from another's.

Continuum 1: Community action to government policy

It is critical to distinguish between the value of landcare groups from a government perspective (the rationale for spending public monies to assist them); the reasons individual land users may have for being in a landcare group and the professed aims of the landcare group.[28]

Figure 7.2: Stewardship is not simple

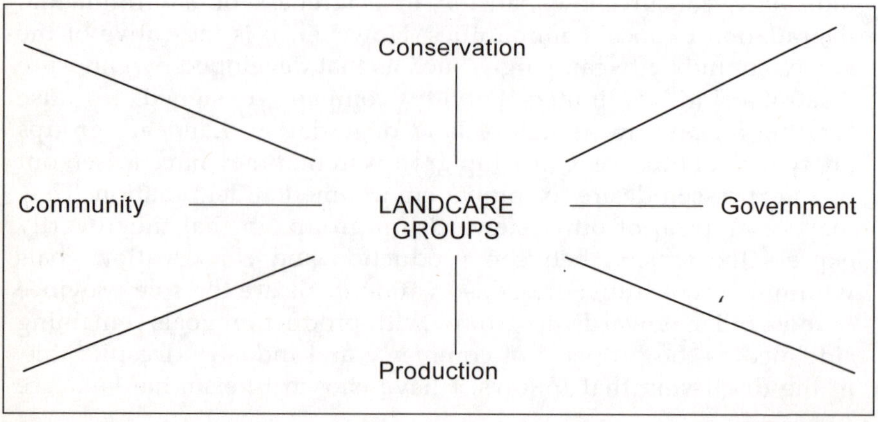

The proponents of community action, "small is beautiful" and populist agendas believe that environmental stewardship as a broad-based social movement will change the face of our global environment through participative processes and bioregional principles. This perspective is strengthened by the convincing argument that rural people the world over have to design their own future and construct solutions to the many faceted problems of land degradation because they have no choice. Advocates of landcare as a community owned, community led revolution in Australian farming systems tout the rise of involvement by women and the increase in "transfer" of agricultural information from one landholder to the next as examples of how this process works at the local level. There are those who think that these groups succeed in attracting government resources to local problems as self-help mechanisms aimed at increasing social equity and self-determination. Others, apparently more cynical, perceive landcare groups as pawns of government policy.

Social theorists wary of the vertical integration of agriculture, cognisant of the increase in agri-businesses, and who have tracked the trend to conservatism in developed nations believe that landcare groups are just another tool of the state. Proponents of this view proclaim that participatory practice is just another way for government to devolve and fragment their environmental responsibilities. They refer to theories of economic rationalism when pointing to landcare groups as unwitting victims of the state, powerless to control their destiny when faced with private interests who want access to their lands and markets and deeper problems of structural inequality in the dominant agricultural development paradigm.[29] For those who look upon local activism as a subtle form of government policy used to lever some land managers off their land in the name of user pays, it is wise not to get caught up in the naive populism espoused by the "community experience" activists.

Continuum 2: Conservation to production

Landcare groups may also be seen as a way of promoting agricultural production goals, through extension services overseen by government officers who are running their own policy agenda. Members of production-oriented landcare groups plant trees – to protect stock and to mitigate the adverse economic effects of dryland salinity, or add amenity value to farms. These group members visit each others' properties to assess different production strategies in different locations. They discuss marketing strategies for different crops, talk of the pros and cons of vertebrate pest management and lament the prohibitive cost of fencing. These landholders perceive that environmental stewardship is too expensive to warrant spending time or money planting trees merely for conservation and prefer to see environmental outcomes as a way of value-adding to private property interests.

However, to other more idealistic environmental commentators, landcare groups can assist the long-term sustainability of Australian agro-ecosystems. In their view groups foster environmental education and nurture green sentiments (although this may be done covertly since to be green is not necessarily a good thing. And to be a "greenie" is almost certainly bad). Yet stewardship activities continue out of concern for a river, a patch of bush, a lost species or a general concern with land and water quality. Community group members of this persuasion plant trees - to save the bush. They collect seed - to ensure survival of tree species. They encourage communication among diverse sections of society. They monitor water quality. They visit field study centres to learn about floodplain ecology. They are the epitome of stewardship itself, acting today in the interests of their environment and of their childrens' future who one day, they hope, will thank them for it.

So where do landcare groups fit?

> Is it soil conservation, good farming, tree planting, wildlife protection, land stewardship, sustainable logging, dune stabilisation, garden composting, mine rehabilitation, farm planning, habitat retention, waste recycling, water quality, land ethics, environmental education, park improvement, catchment planning or land stewardship [sic]?[30]

Landcare groups can encompass all of these activities, and may be comfortable anywhere along the continua of conservation to production, or community concern to government policy. There are a number of other continua which these community groups could also intersect; theory - practice, expert data - local knowledge, pro-active planning - reactive fire fighting. The point is that most of these opposites are not mutually exclusive, and most stewardship groups carry out activities that slide along the spectrum of each continuum at different points in time. To categorise stewardship groups in general as simply touting either end of these seemingly opposing paradigms is too simplistic. One of the most valuable features of group approaches to environmental sustainability is diversity - in strategy, focus, membership and across time.

Emerging themes

As the environment movement develops and changes, so too have the beliefs of the day. Emerging themes include a focus on solutions or, at the least, "do-able" outcomes. At the same time there is increasing pressure to do more with fewer resources as the number of stewardship groups increases. Another emerging theme is the diversity of groups' strategies, approaches and visions for sustainability. But perhaps the most apparent trend is the cry for collaboration, partnership arrangements and consensus processes, in line with the Brundtland report's plea for sustainable development. There is an urgent need for more dialogue and exchange *on an equal footing* between top-down and bottom-up approaches to environmental sustainability. I am not advocating a merger with or cooption of local community stewardship by government agencies. Instead, I am suggesting that effective partnerships for environment stewardship can be implemented in a spirit of trust and cooperation between traditionally separate and distinct agencies. Local decision-makers must negotiate access to resources, opportunities and information from government agencies. I do not use the term "middle-ground" to infer any political reconciliation of the left and the right, or to suggest a position of compromise. It is more a space for pragmatic consideration of how communities and government can support local environmental stewardship. There is more to this type of collaboration than simple compromise, and there are sacrifices to be made on both sides when working toward noticeable improvements in ecological sustainability.

The catchcry still echoing in departmental hallways suggests that community and government should work together, implying the need for a partnership between the community, government and business. One could be cynical and ask why government and community are working together now when they haven't appeared to have had a very productive relationship working on health, education, employment, housing, indigenous affairs or any other major facet of human activity in the past. Or have they? Certainly times have changed recently and there are productive relationships being established between top-down and bottom-up players in the environment game. There are a growing number of authors in the area of agriculture and rural development, for example, who argue that citizens, as part of their local environments, should not only be consulted by governments, they should be active protagonists when creating sustainable alternatives in their region. However, it is appropriate here to sound a cautionary note. Any outsiders' uncritical attachment to participatory processes and practices in an effort to either protect or plunder the environment at the expense of its local stewards and their cultural and institutional arrangements is neither wise nor desirable.[31] Reconsider Figure 7.1 for an overview of the strengths and weaknesses inherent at both top-down or bottom-up scales and for a summary of the integrating links in the middle-ground.

Middle-ground approaches

The middle-ground is about careful listening and clear communication. In essence it is about the philosophy of ecologically sustainable development which unites conservation and development interests, government and non-government organisations and local with global approaches to improving the quality of life on this planet. However, developing sustainability in practice requires a focus on the communicative process which is rarely a feature of the "act now think fast" emphasis on outcomes and products prevalent in modern society. Developing inter-disciplinary, practical and flexible approaches to sustainability requires environmental stewardship – collaborative activities which are focused on making a difference to the landscape. Whether they are called collaborative policy fora, facilitated policy dialogues, cooperative problem-solving techniques, environment round tables, people-oriented techniques, land-use bargaining teams, consensus conferences, citizens' juries or action centred networks, I believe they all require a local territorial contextual base and an emphasis on practical action within a specific landscape. They are characterised by:

- flexibility of membership;
- realistic goals and objectives;
- an ability to respond quickly;
- consensus-based decision-making;

- experiential learning styles;
- non-hierarchical organisational frameworks;
- atmosphere of mutual respect among participants;
- horizontal and vertical links to other networks;
- an openness to restructuring; and
- specific commitments to projects.

The middle-ground could exemplify the best of community and government approaches to environmental management. After all "the basic units of society are not departments, committees, and companies, but rather individual thinking and feeling human beings".[32] The Mitchell River group is a particularly good example of middle ground collaboration. Middle-ground models will flourish as long as attention is paid to the limitations and participatory nature of such approaches, and the rules guiding the principles and practice of collaboration are made overt. General guidelines or recipes do not work. Regional differences based on consideration of sense of community, sense of place, local epistemology and empowerment practice must be reviewed. In short, more attention should be paid to participatory practice in terms of who participates, whose knowledge counts and who should be involved in setting agendas in the middle-ground.

In the environment and sustainable development literature, participatory democracy is emphasised as one way of achieving such collaboration. This is evidenced in *Agenda 21*, the Australian National ESD strategy and *Caring for the Earth*,[33] among a host of other notable publications. The latter listed community empowerment among its principles for building a sustainable society. The full quote is worthy of citing here.

> Most of the creative and productive activities of individuals or groups take place in communities. Communities and citizen's groups provide the most readily accessible means for people to take socially valuable action as well as to express their concerns. Properly mandated, empowered and informed, communities can contribute to decisions that affect them and play an indispensable part in creating a securely based sustainable society.[34]

However, in order for government and community to work together, there are a number of problems in the middle-ground, which need attention relating to the caveat that communities are "properly mandated, empowered and informed". Perhaps one of the more serious errors by government planners is their unerring belief in the infallibility of "community involvement". Just because community groups have the opportunity to become involved in environmental stewardship does not mean that there is automatic problem ownership and development of local solutions. In an Australian review of community planning for salinity management, the authors make the observation that individual landholders may not even recognise dryland salinity. Then, in cases where they do perceive a problem requiring action, there is not an automatic link to a change in behaviour, whether that is in a group or individually. As discussed in Chapter 6, the link between problem identification, attitudes and behaviour is far from linear and is still hotly debated in the social sciences. So, for partnership arrangements to be created in the "middle-ground" which are properly mandated, powerful and well informed, there are a number of potential problems requiring recognition and further discussion.

Problems with the middle-ground

Our natural surroundings in all their beauty, productive capacity, unexpected danger and overwhelming complexity are in danger of experiencing further degradation. Whose problem is that? Whose problem is land degradation, for example? It is a question of ownership. Australian landholders' attitudes would not be unique in stating "my land's my own, to do with what I like". But if there's a problem, then it's government's problem. Some would claim that senior officers within government institutions are responsible for land degradation since they have a detached perspective and more complete information. While recognising the importance attached to this administrative function, it is simply not fair to expect government officers operating within the bonds of bureaucracy to know and understand land degradation in every community. Nor is it desirable. Centralism and the problems of national coordination have helped to create some of the environmental problems which stewardship groups are trying to redress today. Acknowledgement of this has led to the development of regional catchment and basin-wide environmental organisations, eg the Murray-Darling Basin Commission in southeastern Australia. Although coordination and integration functions of regional environmental organisations have been very successful and have had positive impacts on local stewardship, there are dangers in this approach.

With the push for regionalism and the central organisation of environmental issues, there is an element of the "false consciousness" argument, prevalent in Marxist doctrine, for scientists who purport to know and understand the intricacies of the real world "out there" better than local people because the scientists are standing outside that consciousness when making the judgements. It is a little like saying: "I know your reality better than you do and, what is more, you are unable to perceive your hard life because you're stuck in the trap of poverty and rural decline and can't see as clearly as I can". Paternalism was clearly illustrated in Chapter 5 as an attitude of one government agency and constrains the effectiveness of community stewardship. It was especially apparent in the views of government officers in the Mitchell River group who believe that environmental stewardship is an effective (and cheap) way of mobilising a voluntary labour force to carry out actions designed and monitored by government.

From a bottom-up perspective, there are also difficulties in identifying whose problem "it" is, when talking about dryland salinity, soil erosion or water quality to name a few. Typically, facilitators working hard for stewardship groups are placed in a very difficult position if they are torn between loyalty to the group and attachment to the government's mandate by which many keep their job. In the "tiled-drain" example, group members who did not question where the drained water would end up deemed pumping salty water into the river an entirely logical and pragmatic proposition. The group's coordinator was stuck in the middle. Did he support the interests of group members whom he worked for and lived among, or did he think of the bigger picture and the problems that salty water would create for others downstream? What could he do when faced with such a problem of divided loyalties and fragmented subjectivity? At one and the same time he was accountable and loyal to the group and duty bound to the government agency who were paying him and who disapproved of this practice with the wider catchment in mind. Facilitators must alert group members to "the big picture" while simultaneously reminding senior government bureaucrats that it is the local facilitator who is directly responsive to the community.

Environmental problems are constructed in very different ways depending on whose perspective is being considered. On the one hand, stewardship groups feel that they are faced with serious socio-economic and environmental problems to which governments are not attending by virtue of their narrow departmental viewpoints. It is difficult to conceive it otherwise. The structure of the State is, by its very nature, compartmentalised, given that it has to consider a variety of constituencies and manage conflict between them. On the other hand, government departments want to legitimise their activities by proving that communities are accepting responsibility for their own problems and taking the initiative for addressing these at a local level. Yet community groups cannot accept sole responsibility.

> It is simply unfair and totally unrealistic to expect voluntary groups of people, even if there were one hundred thousand active landcare members, to fix land degradation or develop more sustainable farming systems without significant external support – financial, technical, institutional and moral.[35]

When community groups and government departments get together to work on a project as large and complex as environmental degradation, top-down and bottom-up approaches differ tremendously. However, in coming to the middle-ground, agreements are required about some very basic issues such as the geographic scale at which these problems can be addressed, the timeframe in which degradation issues require action, the level and type of resources necessary, the legitimacy or credentials of the people who take these tasks on board and who takes the initiative. Approaches such as multi-stakeholder fora, environment round tables and convenor policy fora trialed in North America can go a long way toward answering some of these questions.

Principles of the middle-ground

In the sense that there are no easy answers to environmental degradation in post-modern society and that we global citizens must be prepared to invent our own solutions, the questions veer away from "whose problem is it?", "how do you fix environmental degradation?" and "how do you know if it is fixed?", toward consideration of how to establish a middle-ground based upon principles of interaction, negotiation and mediation of common goals and conflicting interests. Still, if these questions are ignored altogether then "sustainable damage"[36] will continue to be inflicted upon our environments even if we feel good about it. To restate, I believe that the most important principle for top-down and bottom-up approaches alike is dialogue. To work together and recognise the strengths inherent in each approach requires an organised process of discussion and exchange. However, there are some principles guiding decisions as to who participates in establishing the dialogue, and "when and where participation counts" which reflect a meta-level process concerned with equity and the distribution of power in middle-ground models.

Participatory natural resource management is undergoing a revival as we begin the new millennium. Representative systems of democracy are inequitable in this process. Experts with technical knowledge who represent their department, industry or agency are missing the underlying political point: "even defining the question to be discussed by a collaborative group usually requires sorting out basic choices about social values".[37] Toward that end, nobody is any more or less expert than anyone else. Participatory systems of democracy are much better for middle-ground decision-making – as long as they are open and that all participants understand the language of the debate. In an ideal world they should also participate with equal status". Some would argue that this is a "big ask", however, I maintain that the public involvement of citizens affected by environmental, social and economic decisions in local communities is vital. Before even walking onto the middle-ground, there are some issues concerning public involvement, which need resolution. Decisions must be made about:

- which publics are involved (not yet resolved for the Mitchell River group);
- whose problem is being addressed (of concern in the Water Watchers group);
- who pays the cost of involvement (a potential conflict in the Downside Landcare group);
- what is the object of public involvement (requiring clarification in all three groups).

If we are serious about addressing environmental degradation problems, negotiation and mediation of both the process and content of middle-ground approaches will become increasingly important. And if there is no general consensus about desired directions, there may be no point in proceeding. Dialogue and exchange are essential to this process and can be enacted at the community level through support from State and Federal governments. Actual places where land users and bureaucrats can meet face to face and work on programs of mutual benefit to local catchments are integral to defining the middle-ground. Water

Watchers, the Mitchell River and Downside Landcare groups all met within specific local places, relied upon a blend of local and regional expertise and engaged in collective action, albeit to varying degrees. Perhaps if other groups were encouraged to adopt middle-ground management strategies, more environmental stewardship practice would result in on-ground change.

Conditions for middle-grand environmental management

Based on examples and evidence presented in the preceding chapters, I suggest that cooperative (middle-ground) management arrangements are more likely to succeed if attention is paid to:

- Collaboration between local, historical and traditional forms of knowledge and techno-science.

This is true not only for indigenous communities using traditional ecological knowledge systems, it is also true for farmers and graziers who have established traditional systems of knowledge premised on a commitment to learning by doing, or "action learning". It is necessary to take local ways of knowing into account because they incorporate local history and local information based on day to day experiences, and to combine this with more formal knowledge. The end result is more comprehensive that either system alone.

- Commitment to moving beyond arguments over rights and regulations to the generation of alternate solutions and conflict resolution.

If communities and governments are to share decision-making and cooperate in natural resource management, there needs to be a change in the way that committees and groups are established: away from representative democracy toward more refined participatory democracy. When representatives come to the negotiation table "armed to the teeth" with statements outlining their organisations' positions on issues, the process is not likely to lead to cooperation. Instead, it is well recognised that groups of people who are committed to investigating the issues and exploring all alternatives are more likely to resolve their differences through mediation.

- Facilitation of group dynamics and attention to open administrative arrangements.

Community groups must be aware of the need to create opportunities for informal discussion; establish clear goals and strategies for achieving them; decide upon an appropriate and transparent leadership style; enhance members' opportunities for training and education; establish the right balance of self-reliance and dependence on outside resources; practice clear verbal and written communication; prepare evaluation strategies in advance, and recognise and celebrate when groups meet their goals.

In addition, government and community approaches to environmental stewardship should recognise the threats and opportunities inherent in any local context when seeking complementarity. Group members must have a clear idea of their group's identity and what it is achieving with respect to local conditions and relationships. Cooperative (middle-ground) management can be an effective means of accounting for the complexity in post-modern environments. Here I do not use co-management in its more narrowly defined sense of establishing a system of rights and obligations, rules or procedures for making decisions about single migratory species such as salmon, geese or caribou. Instead co-management can be more broadly defined:

> Co-management is not only about new institutions, but also more fundamentally about the new relationships resulting from them. Institutions and legal arrangements can only permit, support and create incentives for new relationships: It is the new relationships, which generate the communication, trust and willingness to risk innovation which make the benefits of co-management actually materialise.[38]

Model of the middle-ground

The purpose of building this model (Figure 7.3) is to depict the interactive nature of relationships between the principles, profile, processes and products of both top-down and bottom-up environmental management styles and to illustrate the potential of the middle-ground. The model is an *ideal* of top-down and bottom-up approaches to environmental management in the sense that it can be taken apart, reconstructed and moulded to a specific geographic context or to the requirements of the institutions involved. It is one representation and offers explanation of how things are rather than any predictive model of how things should be. The model offers a new way of conceptualising environmental

Figure 7.3: Top-down and bottom-up work toward the middle-ground

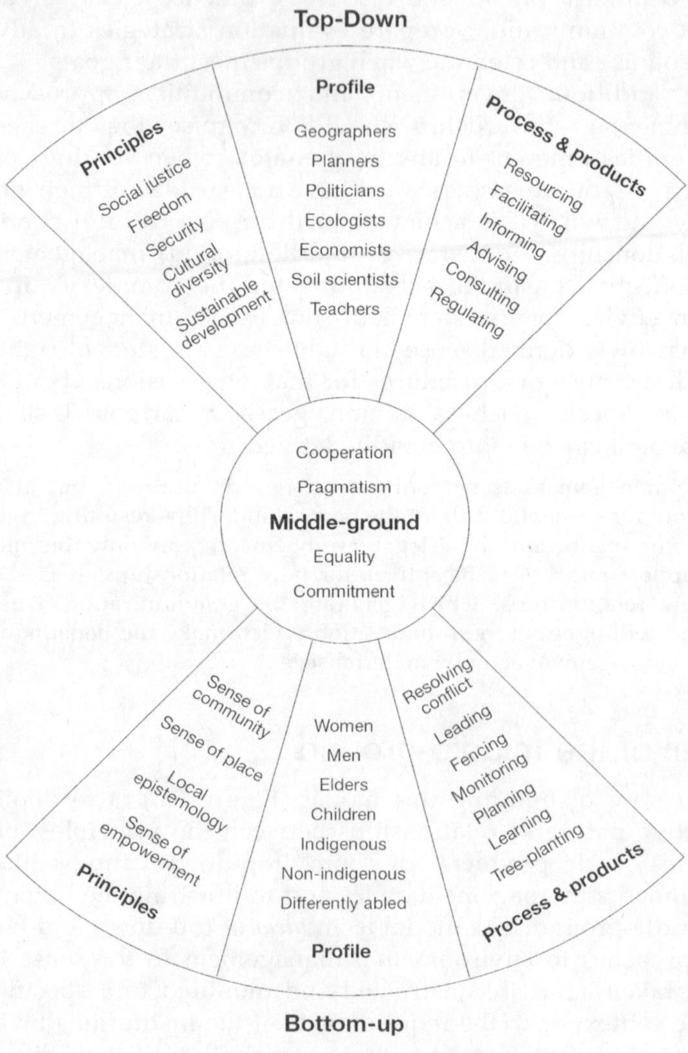

stewardship, and provides a way forward in analysing and improving community involvement in environmental management. However, like other models, it is simplistic in the sense that the relative importance of each category is context dependent and fails to account for human volition. While each stewardship group does have a "sense of community", the extent to which it affects that group's sense of empowerment differs from group to group.

The greatest asset of a model of this kind is its potential for offering a new way of thinking about or conceptualising the issues. On the other hand, it could be a liability if interpreted within an overly reductionist methodological framework. The hour-glass shape of the model does reflect a capacity for movement and flow, but should not be interpreted as any attempt to reinforce the hierarchical relationships implicit within its vertical orientation. Rather than perceiving government agencies as the benevolent patrons of sustainability, issuing resources and advice from a position "on high" of perfect knowledge and informed democracy, one can conceive of a government system which thrives on the grass-roots input of local communities. Or, to put it the other way around; rather than thinking of community groups as merely the passive instruments through which government environmental policy is enacted from above, one can turn the hour-glass upside down and think of powerful, thriving and energetic stewardship groups operating even more effectively when government agencies can offer support such as advice and resources specifically tailored to local needs. The line separating the middle-ground from both the top-down and bottom-up is not static. Both ends can (and should) move back and forth to the middle.

Again, it is important to stress that cooperation in the middle-ground does not necessarily equal compromise between the top-down and bottom-up. Nor does it equate to "lowest common denominator" environmental stewardship. As noted before, there is more of a continuum of approaches some of which will be more appropriate in specific circumstances than others. Although compromise is one way of approaching the middle-ground, conflict over important principles is an equally valid approach, and sometimes mediation and arbitration may be required. Communication, not compromise, is a more effective management tool in the middle-ground:

> Formulating value positions that are broad and inclusive is a delicate task; it requires considerable diplomacy and an

> exchange of views that does not attribute dogmatic authority to anyone.[39]

"Narrow" science communication efforts as well as broad ones are required in order to appeal to public and specialised audiences. Environmental stewardship is promoted if people are provided with the means of making their own choices, the grounds for making informed choices and the institutional arrangements that encourage public involvement in making choices. Although this is an ideal situation, it is far from becoming a reality for many environmental stewards. Many critical issues and outstanding questions require discussion and classification including:

> Can institutions be designed that would be conducive to a genuinely inclusive politics? How can they avoid favouring certain kinds of participants, reflecting a situated and privileged view of the way the world works? Perhaps more importantly, will those promising expanded opportunities for public participation be concerned about selective or biased inclusiveness?[40]

Middle-ground approaches can be examples of such institutions, but are doomed to failure if they opt for representative "stakeholder" systems of democracy aimed only at achieving "balanced inclusiveness". Through the creation of new ways of acting politically and new spaces for political action, the situated and diverse positions of the poor, marginalised, indigenous and other minorities will be empowered to contribute. The middle ground, I think, should be characterised by political action which aims "at arousing aspiration, raising consciousness, exploring the politics of identity, and opening up opportunities for those who are marginal".[41] The middle-ground seems a fitting context for such goals.

Middle-ground environmental stewardship is premised upon:

- equity and diversity among horizontal and vertical links,
- a commitment to seeking solutions and openness to new ideas; and
- flexibility, responsiveness and pragmatism.

The middle-ground would become more realistic than merely potential if both top-down and bottom-up environmental managers paid greater attention to these themes of commitment, equality and pragmatism. A commitment to stay in the process, while at the same time remaining open to new ideas is important for environmental stewardship. Based on a recognition and respect for the practices of both governments and communities, commitment to principles such as a sense of community and freedom of expression may first appear to be diametrically opposed, but fundamentally reflect a concern for both the needs of the many and the needs of the few. Moving into the centre column of the model, equity and diversity may be expressed through indigenous people being economists, children being teachers, women being ecologists and left-wing political supporters working alongside advocates of right-wing philosophies. In other words, middle-ground environmental stewards may represent a variety of ideological positions and interests while at the same time engaging in issue-specific decisions relating to stewardship projects. Lastly, pragmatism in processes and products at either end of the top-down to bottom-up continuum can result in integrated efforts, negotiated outcomes and demonstrable achievements. In this category, it is important to note that resolving conflict is just as important as planting trees for community groups. Importantly, legislation does not assume any greater importance than consultation on behalf of government agencies engaged in environmental "solutions" – they are both valid methodological tools to bring to the middle-ground.

There are still many questions to be answered about how middle-ground models relate to community and government approaches to environmental stewardship. For example:

- Where do middle-ground models fit into the current move toward regionalisation and structural adjustment?
- What types of middle-ground models currently exist internationally? How can they be better funded, administered and coordinated?
- What are the information and resource requirements of middle-ground models? How are these currently met? To whom should these resources be targeted?

Summary

The problems of working toward the middle-ground are tremendous. Even when all stewardship players have made a commitment to work together and have addressed some of the problems identified above, there are other difficulties. After having decided, for example, on whose problem environmental degradation is and who participates in its resolution, there are other questions as to who pays for it and how to evaluate it. In the current climate of economic rationality, it is very appealing to dwell upon questions of accountability, evaluation and monitoring. Despite wishful thinking that: "our problem is not one of economics, but one of estranged values, where our economics is divorced from ethics",[42] I don't believe that this argument is one that government financial advisers and auditors of public spending are ready to hear yet. There have been too many changes to environmental policy frameworks over too short a space of time. Not least of these changes is recognition that environmental stewardship groups are providing the whole of society with a means of addressing issues of environmental degradation. Some of the more fundamental changes are illustrated in Table 7.1.

Table 7.1: Local environmental management: changing principles for action

Changes From	Changes To
Short-term focus, politically driven	Long-term focus, defined by whole community
Perceived as technical problem requiring technical solutions	Perceived as symptomatic of wider socio-economic, cultural and biophysical problems
Extension science engaged to educate the people about the environment	Action research approach used for government and community to learn together
Environmental "science" is a male dominated issue	Environmental "studies" concern everyone
Outcome-driven, goal-oriented	Emphasis placed equally on process and outcome, means and ends
Measurements of success evaluated in terms of on-ground change quantifiable at the bottom of the catchment	Success measured in less quantifiable terms, recognition of less tangible indicators of success

Middle-ground environmental stewardship implies an integrated approach to natural resource issues, but it goes deeper than merely throwing all the necessary ingredients together and trusting that group process will look after itself. Middle-ground approaches also act to bridge private-public sector rifts and link individual households with wider society through the establishment of local stewardship groups. They are not only transdisciplinary, but also cut across departmental barriers and frequently attempt the seldom traversed institutional divide of government-community relations. Finally, local environmental groups are approaching the often espoused political slogan "community and government working together" not through blindly adopting government policy, but through a process of transactive planning, communication and cooperative stewardship.

- There are no universal answers to local environmental problems. Given the complexities and uncertainties of unsustainable environments, *local* attempts to find *local* solutions are not just a current trend, but crucial to establishing management regimes that are effective in particular contexts.
- Community groups and government agencies must embark on a course of middle-ground environmentalism and encourage cooperative management of local places. This necessitates open dialogue and exchange to allow communication in planning, and action toward "solving" local environmental problems.

In conclusion, environmental stewardship is critically important for sustainability, providing that local groups have the necessary financial and educational support information and human resources to manage their local environments. This includes proper legal mandates, effective consultative processes and a state-sanctioned vision of sustainability. However, these are necessary but not sufficient principles with which to build sustainable environments. A micro-scale focus is also required, one which is place-centred and includes a sense of community, opportunities for empowerment and attention to local ways of knowing.

The world is still watching to see if stewardship groups are working effectively to combat environmental degradation. And rural crises the world over are still very big problems to which many people are seeking local solutions. Environmental stewardship groups are not *the* answer, but they are a good way of addressing problems of unsustainable development. In the same way that there is more than one problem, so there is more than one way of tackling environmental problems. What is missing from our language about sustainable environmental management is balance. The needs of the few (person-in-community, the group) must be measured alongside the needs of the many (broader society, "the common good') in order to avoid isolation and fragmentation and to protect commonly held natural resources. Paradoxically there is a difference between governments' expectations of how stewardship groups work and what the groups produce as a result. And there is another difference between the community's expectations of government and government's own bureaucratic processes and approaches. In both there is an emphasis on attaining results (products) which skims over the process involved. Now that large institutions are attending to these decision-making styles with more rigour, there should also be an attendant emphasis on how stewardship groups work. Most importantly, there needs to be more energy spent on both community and government's behalf toward fulfilling human needs, maintaining ecological integrity, achieving equity and providing for systems of social self-determination in the middle-ground.

If government agencies and community groups are working cooperatively toward sustainability, then it is appropriate that both top-down and bottom-up approaches are used when negotiating the middle-ground. If, however, there are tensions in attitudes and values around establishing a common discourse between government and community, it will become apparent that local environmental stewardship is not an uncontested goal even though it may have become a universal practice.

> It may well be that wild-flowers grow by themselves. But grass-roots organisations do not. They are cultivated, in large measure, by just policies and competent government agencies that do their job.[43]

I hope that the experience of the Water Watchers, Downside Landcare and Mitchell River groups will inspire others to support environmental stewardship groups and learn from them as I have done.

Endnotes

1 In the midst of another bout of "science-wars", I can only affirm my belief that scholars of science (those conducting studies of the theory and practice of science - not usually practising scientists) should not all be labelled anti-science simply because they take a critical theory perspective. Nor should it be assumed that they do not take nature seriously because they use methods from the social sciences and humanities and "give ontological and epistemic primacy to language and social practices" (*http: //www.darkwing.uoregon.edu/~tns*).

2 Pretty and Chambers 1994: 198.

3 Wynne 1989, 1992, 1995, 1996a; Irwin and Wynne 1996; Jasanoff 1997; McDonell 1997, Yearley 1994.

4 See the Loka Institute Community Research Network (*http://www.loka.org/crn*); Fixdal 1997 and Joss 1998 on consensus conferences.

5 Paper delivered at the 1999 International Symposium on Society and Resource Management by Mark Brunson of Utah State University.

6 Campbell 1994: 16.

7 For a more detailed discussion of the roles of scientists in environmental decision-making, see Harding 1998.

8 Carley and Christie 1992: 160.

9 Alan Cottey points out that although freedom of information (or, more accurately freedom from secrecy) may have become universally attractive, not all secrecy is "to be opposed indiscriminately". He also suggests that "different kinds of openness are appropriate and achievable in different areas of cultural activity such as commerce, law, government, politics, administration, sport, applied science and basic science" (Cottey 1998: 89)

10 Papadakis 1996: 26.

11 Campbell 1992a; Martin, Tarr and Lockie 1992; Beilin, 2001.

12 Brunson 1999: 2.

13 Farley and Toyne 1994, 2000.

14 All these changes took place between 1983 and 1995, four changes in twelve years. It is little wonder that in 1992 landcare stewards in Victoria told me that the CNR was colloquially known as the Department for Constant Name Revision. A colleague, Libby Robin, informed me later that it was also known as the Department which Can Not Remember!

15 Langton 1982.

16 Powell 1976: 173.

17 M'Gonigle 1986: 305.

18 Beilin 2000; Beilin 2001.

19 Gray 1991, 1992; Wild 1974; Dempsey 1990; Gray, Dunn and Phillips 1997.

20 Bryant and Preston 1987: 4.

21 See King 1994; , Fraser 1994 for clarification and description of rural suicides. Arguably this is one extreme example of rural misery.
22 Common Law Mortgage 1924, provided by a Downside Landcare group member.
23 Auvine, Densmore, Extrom, Poole and Shanklin 1977; Dick 1991; West 1992; Chamala and Keith 1995.
24 Olsen 1965: 52.
25 Curtis 2000: 57.
26 See Kloppenburg 1991; Scoones and Thompson 1994; Pretty and Chambers 1994; Chambers 1997; Millar and Curtis 1999.
27 Lane, Dunn and Cregan 1994.
28 Campbell 1992a: 13.
29 Furze 1992; Lockie 1993.
30 Roberts 1992: 3.
31 Consider some international community development agencies often clumsy attempts to persuade local people to "participate" in conservation efforts within third world contexts (Filer n.d.).
32 Trzyna 1990: 10.
33 United Nations 1992; Commonwealth of Australia 1992; IUCN/UNEP/WWF 1991.
34 IUCN/UNEP/WWF 1991: 11.
35 Campbell 1992a: iii; Curtis 2000 says the same thing eight years later.
36 Gammage 1994.
37 Trzyna 1990: 12.
38 Pinkerton 1989: 8. The term "collaborative management" of protected areas refers to a partnership by which stakeholders agree on sharing among themselves the management functions, rights and responsibilities for a territory or set of resources under protected status. The stakeholders primarily include the agency in charge and various associations of local residents and resource users, but can also involve non-governmental organisations, local administrations, traditional authorities, research institutions, businesses, and others Borrini- Feyerabend 1996: 3.
39 Rosenau 1992: 146.
40 Schrecker 1994: 36.
41 Rosenau 1992: 147.
42 Nozick 1992: xi.
43 Annis 1987: 133.

References

Agriculture and Resource Management Council of Australia and New Zealand, Australian and New Zealand Environment and Conservation Council Forestry Ministers, 1997, *The National Weeds Strategy: A Strategic Approach to Weed Problems of National Significance*, Australian Government Publishing Service: Canberra.

Alexander, D, 1990, "Bioregionalism: science or sensibility?" *Environmental Ethics* 12 (Summer, 1990): 161-173.

Alexandra, J, Haffenden, S and White, T, 1996, *Listening to the Land: A Directory of Community Environmental Monitoring Groups in Australia*, Canberra: Australian Conservation Foundation.

Annis, S, 1987, "Can small-scale development be a large-scale policy? The case of Latin America", *World Development* 15, Supplement: 129-134.

Anon, 1994, "A profile of volunteer monitoring", *The Volunteer Monitor* 6(1): 4-5.

Arnstein, SR, 1969, "A ladder of citizen participation", *Journal of American Institute of Planners* 35: 216-224.

Australian Bureau of Statistics, 1999, *Environmental Issues: People's Views and Practices*, Australian Bureau of Statistics: Canberra, (ABS Catalogue No 4602.0).

Auvine, B, Densmore, B, Extrom, M, Poole, S and Shanklin, M, 1977, *A Manual for Group Facilitators*, Centre for Conflict Resolution: Madison.

Barr, N and Wilkinson, R, 1992, *Community Planning in the Victorian Salinity Program: An Evaluation*, Department of Food and Agriculture: Melbourne.

Bawden, R, 1991, "Reflections on the action of reflections in action: of cars, helicopters and satellites", in Zuber-Skerritt, O (ed) *Action Learning for Improved Performance*, AEBIS Publishing: Brisbane.

Beilin, R, 2001, "The brave new order: power, visibility and the everyday landscapes of Australian farmers", in Higgins, V, Lawrence, G and Lockie, S (eds), *Social Sciences in Natural Resource Management: Theoretical Perspectives*, Edward Elgar: London.

Beilin, R, 2000, "Sustaining a vision: recognising landscape futures", Paper delivered to International Landcare 2000 Conference, Melbourne, 2-5 March 2000.

Blaikie, P, Brown, K, Stocking, M, Tang, L, Dixon, P and Sillitoe, P, 1997, "Knowledge in action: local knowledge as a development resource and barriers to its incorporation in natural resource research and development", *Agricultural Systems* 55(2): 217-237.

Blakely, ED, 1989, "Theoretical approaches for a global community", in Christenson, JE and Robinson (Jr), JW (eds), 1898, *Community Development in Perspective*, Iowa State University Press: Ames.

Borrini-Feyerabend, G, 1996 (ed), *Collaborative Management of Protected Areas: Tailoring the Approach to the Context*, IUCN – The World Conservation Union: Gland Switzerland.

Botelho, Z, 1999, "Once and future creeks: plans for stream daylighting promise resurrection of Vancouver waterways", *Alternatives* 25 (2): 6.

Bradby, K, 1992, "The art of public listening – stories from the Peel-Harvey", in Greening Australia (ed) *Catchments of Green Conference Proceedings Volume A*, Greening Australia: Canberra.

Braysher, M, 1993, *Managing Vertebrate Pests: Principles and Strategies*, Australian Government Publishing Service: Canberra.

Brunson, MW, 1999, "Observing vs doing: a researcher learns about collaboration", Paper delivered to the 1999 international Symposium on Society and Resource Management, University of Queensland, Brisbane, Australia: 7-10 July 1999.

Bryant, CR and Preston, RE, 1987, "A framework for local initiatives in economic development", *Economic Development Bulletin 1*, Economic Development Program, University of Waterloo: Waterloo.

Buchy, M and Ross, H, 2000, *Enhancing the Information Base on Participatory Approaches in Australian Natural Resource Management*, LWRRDC: Canberra.

Buller H and Hoggart, K, 1986, "Nondecision-making and community power – residential control in rural areas", *Progress in Planning* 25: 131-203.

Burton, J, 1992, "Catchment management in Australia – an historical review", in Greening Australia (ed) *Catchments of Green Conference Proceedings Volume A*, Greening Australia: Canberra.

Campbell, A, 1992a, *Landcare in Australia: Taking the Long View in Tough Times*, National Soil Conservation Program, Department of Primary industries and Energy: Canberra.

Campbell, A, 1992b, "Community participation – the 'catch' in catchments", in Greening Australia (ed) *Catchments of Green Conference Proceedings Volume A*, Greening Australia: Canberra.

Campbell, A, 1994, *Community First: Landcare in Australia*, international institute for Environment and Development: London, (Gatekeeper Series No, SA42).

Campbell, A and Junor, R, 1992, "Land management extension in the '90s: Evolution or emasculation?" *Australian Journal of Soil and Water Conservation* 5 (2): 16-23.

Carley, M and Christie, I, 1992, *Managing Sustainable Development*, Earthscan: London.

Carr, AJL, 1992, *Community Participation in Water Quality Monitoring: The Case of Water Watchers*, Centre for Resource and Environmental Studies, Australian National University: Canberra, (CRES Working Paper 1992/6).

Carr, AJL, 1993a, *Community Involvement in Landcare: The Case of Downside*, Centre for Resource and Environmental Studies, Australian National University: Canberra, (CRES Working Paper 1993/2).

Carr, AJL, 1993b, *Catchment Co-management: The Case of the Mitchell River Watershed Management Working Group*, Centre for Resource and Environmental Studies, Australian National University: Canberra, (CRES Working Paper 1993/7).

Carroll, J, 1982, Preface, in Carroll, J (ed), *Intruders in the Bush: The Australian Quest for Identity*, Oxford University Press: Melbourne.

Cary, J, Wilkinson, R, Barr, N and Milne, G, 1993, "Establishing the basis for effective care of rural land", *Australian Journal of Soil and Water Conservation*, 6(1): 44-49.

Chamala, S and Keith, K (eds), 1995, *Participative Approaches for Landcare: Perspectives, Policies and Programs*, Australian Academic Press: Brisbane.

Chamala, S and Mortiss, PD, 1990, *Working Together for Land Care: Group Management Skills and Strategies*, Australian Academic Press: Brisbane.

Chambers, R, 1997, *Whose Reality Counts: Putting the Last First*, Intermediate Technology Publications: London.

Checkoway, B, 1995, "Six strategies for community change", *Community Development Journal* 30 (1): 2-20.

Christenson, JE and Robinson, JW, 1989, "Themes of community development", in Christenson, JE and Robinson (Jr), (eds), *Community Development in Perspective*, Iowa State University Press: Ames

Christenson, JE and Robinson (Jr), (eds), *Community Development in Perspective*, Iowa State University Press: Ames

Commonwealth of Australia, 1992, *National Strategy for Ecologically Sustainable Development*, Australian Government Publishing Service: Canberra.

Condon, RW, 1991, "Affordability: the key", *Curlew* 3 (Winter, 1991): 3, (Broadsheet published by the Murray-Darling Basin Commission, Canberra).

Cottey, A, 1998, "Open science and practicable politics", in Star, C (ed) *Green Politics in Grey Times*, University of Melbourne: Melbourne (Ecopolitics XI Conference Proceedings): 88-96.

Council of Australian Governments, 1992, *Inter-governmental Agreement on the Environment*, Council of Australian Governments: Canberra.

Crombie, A, 1995, "Adult learning in group environments: the hidden curriculum", in Chamala, S and Keith, K (eds), *Participative Approaches for Landcare: Perspectives, Policies and Programs*, Australian Academic Press: Brisbane.

Crook, S and Pakulski, J, 1995, "Shades of green: public opinion on environmental issues in Australia", *Australian Journal of Political Science* 300: 39-55.

Curtin, J, 1999, "New public management meets civic discontent", *Australian Journal of Public Administration*, 59 (1) March 2000: 115-124.

Curtis, A, 2000, "Exploring Landcare from a volunteer perspective", *Rural Society* 10(1): 47-61.

Dalton, JH, Elias, MJ and Wandersman, A, 2001, *Community Psychology: Linking Individuals and Communities*, Wadsworth Thomson Learning: Stamford.

Davidson, OG, 1993, "What exactly is community?" *Utne Reader* (60) (November/December, 1993).

Dempsey, K, 1990, *Smalltown: A Study of Social Inequality, Cohesion and Belonging*, Oxford University Press: Melbourne.

Department of Agriculture, 1991, *Decade of Landcare: Interim Plan for Western Australia*, Department of Agriculture: Perth.

Department of Home Affairs and Environment, 1984, *National Conservation Strategy for Australia*, Australian Government Publishing Service: Canberra.

Dick, B, 1991, *Helping Groups to be Effective*, Interchange: Chapel Hill.

Dovers, S, 1994, "Policy processes for sustainability", Centre for Resource and Environmental Studies, Australian National University: Canberra, (Mimeo).

Dovers, S, 1995, "Information, sustainability and policy", *Australian Journal of Environmental Management* 2(3): 142-156.

Dovers, SR and Handmer, JW, 1995, "Ignorance, the precautionary principle and sustainability", *Ambio* 25: 92-97.

Dovers, SR and Lindenmayer DB, 1997, "Managing the environment: rhetoric, policy and reality", *Australian Journal of Public Administration* 56(2).

Esteva, G, 1987, "Regenerating people's space", *Alternatives* XII: 125-152.

Evans, G and Cohen, S, 1987, "Environmental stress", in Stokols, D and Altman, I (eds), *Handbook of Environmental Psychology*, John Wiley & Sons: New York, (Volume 1).

Farley, R and Toyne, P, 1994, "Is landcare working?" Geography 3010: Environmental policy and planning, Australian National University: Canberra, (Guest lecture).

Farley, R and Toyne, P, 2000, "Landcare – looking backwards and looking forwards", Address to the National Press Club, Canberra, 26 July 2000, (Speech notes).

Ferguson, W, 1999, "Fixing a salt marsh: citizens, shovels and sweat", *The Volunteer Monitor* 11(1): 5-6

Filer, C, nd, "How can western conservationists talk to Melanesian landowners about indigenous knowledge?" (Seminar paper) Resource Management in Asia-Pacific Project, Research School of Pacific and Asian Studies, Australian National University, Canberra: 3 February 2000.

Fixdal, J, 1997, "Consensus conferences as 'extended peer groups'", *Science and Public Policy* 24(6): 366-376.

Flores, D, 1999, "Place: thinking about bioregional history", in McGinnis, MV (ed), *Bioregionalism*, Routledge: London.

Fraser, M, 1994, "What about us? The legacy of suicide", *Rural Society* 4(3/4): 7-10.

Frawley, K, 1992, "A 'green' vision: the evolution of Australian environmentalism", in Anderson, K and Gale, F (eds) *Inventing Places: Studies in Cultural Geography*, Longman Cheshire: Melbourne.

Friedmann, J, 1992, *Empowerment: the Politics of Alternative Development*, Blackwell: Cambridge.

Friere, P, 1975, *Pedagogy of the Oppressed*, Penguin Educational: Harmondsworth (paperback edition).

Funtowicz, SO and Ravetz, JR, 1991, "A new scientific methodology for global enviromental issues", in Costanza, R (ed) *Ecological Economics: the Science and Management of Sustainability*, Columbia University Press: New York.

Furze, B, 1992, "Environmental management and capitalist agriculture", in Lawrence, G, Vanclay, F and Furze, B (eds), *Agriculture, Environment and Society*, MacMillan: South Melbourne.

Gammage, B, 1994, "Sustainable damage: the environment and the future", in Dovers, S (ed), *Australian Environmental History: Essays and Cases*, Oxford University Press: Melbourne.

Gardner, JE, 1991a, "Environmental non-government organisations and management of water resources in the Faser River basin", in Dorcey, AHJ and Griggs, JR (eds), *Water in Sustainable Development: Exploring our Common Future in the Faser River Basin*, Westwater Research Centre, University of British Columbia: Vancouver.

Gardner, JE, 1991b, "Environmental non-government organisations and the management of the aquatic environment for sustainable development", in Dorcey, AHJ and Griggs, JR (eds), *Perspectives on Sustainable Development in Water Management: Towards Agreement in the Faser River Basin*, Westwater Research Centre, University of British Columbia: Vancouver.

Gardner, JE and Roseland, M, 1989, "Acting locally: community strategies for equitable sustainable development", *Alternatives* 18 (3): 36-48.

Garkovich, LE, 1989, "Local organisations and leadership in community development", in Christenson, JE and Robinson (Jr), JW (eds), *Community Development in Perspective*, Iowa State University Press: Ames.

Gathcart-Borer, M and Smart, R, 1947, *Bush Christmas*, Sir Issac Pitman & Sons: Melbourne.

Ghai, D and Vivian, JM, 1992, "Introduction", in Ghai, D and Vivian, JM (eds), *Grass-roots Environmental Action: People's Participation in Sustainable Development*, Routledge: London.

Government of Victoria, 1988, *Salt Action: Joint Action: Victoria's Strategy for Managing Land and Water Salinity*, Salinity Bureau: Melbourne.

Grabosky, PN, 1994, "Indirect environmental regulation through green commerce", Centre for Resource and Environmental Studies, Australian National University: Canberra, (Seminar paper).

Grabosky, P and Gant, F, 1999, *Improving Environmental Performance, Preventing Environmental Crime*, Australian Institute of Criminology: Canberra.

Graumann, CF and Kruse, L, 1990, "The environment: social construction and pyschological problems", in Himmelweit, HT and Gaskell, G (eds), *Societal Psychology*, Sage Publications: Newbury Park (Sage Focus Edition).

Gray, I, 1991, *Politics in Place: Social Power Relations in an Australian Country Town*, Cambridge University Press: Cambridge.

Gray, I, 1992, "Power relations in rural communities: implications for environmental management", in Lawrence, G, Vanclay, F and Furze, B (eds), *Agriculture, environment and society*, MacMillan: South Melbourne.

Gray, I, Dunn, T and Phillips, E, 1997, "Power, interests and the extension of sustainable agriculture", *Sociologia Ruralis* 37: (1)

Grierson, M, 1993, "Consumer participation in management of comunity-based social welfare services", in Centre for Australian Community Organisations and Management (ed) *Power, Politics and Performance: Community Management in the 90s*, The Centre for Australian Community Organisations and Management, University of Technology: Sydney.

Hamilton, C and Throsby, D (eds), 1998, *The ESD Process: Evaluating a Policy Experiment*, Academy of the Social Sciences in Australia and Graduate Program in Public Policy, Australian National University: Canberra.

Harding, R (ed) 1998, *Environmental Decision-Making: The Roles of Scientists, Engineers and the Public*, The Federation Press: Sydney.

Hanson, HC, 1962, *Dictionary of Ecology*, Philosophical library: New York.

Heller, K, 1989, "The return to community", *American Journal of Community Psychology* 17 (1): 1-15.

Heller, K, 1990, "Social and community intervention", *Annual Review of Psychology* 41: 141-168.

Hillery, GA, 1955, "Defintions of community: areas of agreement", *Rural Sociology* 20: 111-123.

Hobfoll, SE, Briggs, S and Wells, J, 1995, "Community stress and resources: actions and reactions", in Hobfoll, SE and de Vries, MW (eds), *Extreme Stress and Communities: Impact and Intervention*, Kluwer Academic Publishers: Maastricht.

Hollick, M and Mitchell, B, 1991, *Integrated Catchment Management in Western Australia: Background and Alternative Approaches*, Centre for Water Research, University of Western Australia: Nedlands.

Hunn, E, 1993, "What is traditional ecological knowledge?" in Williams, NM and Baines, G (eds), *Traditional Ecological Knowledge: Wisdom for Sustainable Development*, Centre for Resource and Environmental Studies, Australian National University: Canberra.

Ife, J, 1995, *Community Development: Creating Community Alternatives – Vision, Analysis and Practice*, Addison Wesley Longman: Melbourne.

Irwin, A and Wynne, B (eds), 1996, *Misunderstood Science? The Public Reconstruction of Science and Technology*, Cambridge University Press: Cambridge.

Ittelson, WH, Proshansky, HM, Rivlin, LG and Winkel, GH, 1974, *An Introduction to Environmental Psychology*, Holt, Rinehart and Winston, Inc: New York.

IUCN/UNEP/WWF, 1991, *Caring for the Earth: A Strategy for Sustainable Living*, Gland: Switzerland.

Jasanoff, S, 1997, "Civilisation and madness: the great BSE scare of 1996", *Public Understanding of Science* 6: 221-232.

Johnson, K, 1994, "Creating place and landscape", in Dovers, S (ed) *Australian Environmental History: Essays and Cases*, Oxford University Press: Melbourne.

Joss, S, 1998, "Danish consensus conferences as a model of participatory technology assessment: an impact study of consensus conferences on Danish Parliament and Danish public debate", *Science and Public Policy* 25(1): 2-22.

Karnoouh, C, 1986, "The lost paradise of regionalism: the crisis of post-modernity in France", *Telos* 67: 11-26.

Kaye, G and Wolff, T (eds), 1997, *From the Ground Up: A Workbook on Coalition Building and Community Development*, AHEC/Community Partners: Amherst.

King, R, 1994, "Suicide prevention: dilemmas and some solutions," *Rural Society* 4(3/4): 2-6.

Kloppenburg (Jr), J, 1991, "Social theory and the de/reconstruction of agricultural science: local knowledge for an alternative agriculture", *Rural Sociology* 56(4): 519-548.

Knight, RL and Landres, PB (eds), 1998 *Stewardship Across Boundaries*, Island Press: Washington DC.

Korten, DC, 1980, "Community organisation and rural development: a learning process approach", *Public Administration Review* (September – October, 1980: 480-511.

Korten, DC, 1990, *Voluntary Organizations and the Challenge of Sustainable Development*, Australian Development Studies Network: Canberra, (Briefing Paper # 15).

Kowanyama Aboriginal Land and Natural Resource Management Office and Kowanyama State High School, 1993, *Kowanyama Curriculum Development Project Proposal*, Kowanyama Aboriginal Land and Natural Resource Management Office and Kowanyama State High School: Kowanyama.

Kretzmann, JP and McKnight, JL, 1993, *Building Communities from the Inside Out: A Path Toward Finding and Mobilizing a Community's Assets,* ACTA Publications: Chicago.

Lappe, FM and DuBois, PM, 1994, *The Quickening of America: Rebuilding our Nation, Remaking our Lives,* Jossey-Bass: San Francisco.

Lane, J, Dunn, T and Cregan, P, 1994, *The Kyeamba Valley: Developments in Landcare,* Centre for Conservation Farming, Charles Sturt University: Wagga Wagga, (Mimeo).

Langton, S, 1982, "The new voluntarism", in Harman, JD, *Voluntarism in the Eighties: Fundamental Issues in Voluntary Action,* University Press of America: Washington.

Lawrence, G, 2000, "Global perspectives on rural communities - trends and patterns", Paper No 188.00, International Landcare 2000 Conference, Melbourne, 2-5 March 2000.

Leopold, A, 1966, *A Sand County Almanac: With other Essays on Conservation from Round River,* Oxford University Press: New York.

Leinberger, P and Tucker, B, 1992, "The boomers' search for community: is it true community or tribal networking?" *Utne Reader* 52: 86-88 (July/August, 1992).

Lerner, S, 1993, "Environmental stewardship: a field coming into focus", in Lerner, S (ed) *Environmental Stewardship: Studies in Active Earthkeeping,* Department of Geography, University of Waterloo: Waterloo, (Department of Geography Publication Series No 39).

Lerner, S, 1994, "Local stewardship: training ground for an environmental vanguard", *Alternatives* 20 (2): 14-19.

Levine, M and Perkins, D, 1987, *Principles of Community Psychology: Perspectives and Applications,* Oxford University Press: New York.

Lewis, HT, 1993, "Traditional ecological knowledge - some definitions", in Williams, NM and Baines, G (eds), *Traditional Ecological Knowledge: Wisdom for Sustainable Development,* Centre for Resource and Environmental Studies, The Australian National University: Canberra.

Littrell, DW and Hobbs, D, 1989, "The self-help approach", in Christenson, JE and Robinson (Jr), JW (eds), *Community Development in Perspective,* Iowa State University Press: Ames.

Lockie, S, 1993, "Landcare, self-help, local knowledge and the State", in Jolly, B and Holland, I (eds), *Facing the Future: Ecopolitics VII Conference Proceedings,* Faculty of Environmental Sciences, Griffith University: Brisbane.

Lockie, S and Vanclay, F, 1997, *Critical Landcare,* Centre for Rural Social Research, Charles Sturt University: Wagga Wagga.

Lothian, JA, 1994, "Attitudes towards the environment: 1975 to 1994", *Australian Journal of Environmental Management* 1(2): 78-79.

Mahanty, S, Guijt, I and Carr, A, (comps), 1997, *Tracking Change: Community Monitoring in Natural Resource Management,* Workshop proceedings, Centre for Resource and Environment Studies, Australian National University, (mimeo).

Martin, P and Lockie, S, 1993, "Environmental information for total catchment management: incorporating local knowledge", *The Australian Geographer* 24 (11): 75-85.

Martin, P, Pinn, J, Woodhill, J and Wilson, AM, 1992, "Perspectives and approaches for community participation in environmental care", in Hamilton, GJ, Howes, K,M and Attwater, R (eds), *Proceedings of the 5th Australian Soil Conservation Conference,* Department of Agriculture: Perth, (Volume 9: Community Awareness and Education Workshop).

Martin, P, Tarr, S and Lockie, S, 1992, "Participatory environmental management in New South Wales: policy and practice", in Lawrence, G, Vanclay, F and Furze, B (eds), *Agriculture, Environment and Society,* MacMillan: South Melbourne.

Martin, P and Woodhill, J, 1995, "Landcare in the balance: government roles and policy issues in sustaining rural environments", *Australian Journal of Environmental Management* 2: 173-83.

McDonell, G, 1997, "Scientific and everyday knowledge: trust and the politics of environmental initiatives", *Social Studies of Science* 27(1997): 819-863.

McEachern, D, 1993, "An ideal polity and the possibility of ecological wellbeing", Paper delivered to the conference on individual and Social Well-being, The Environmental Management and Development Program, National Centre for Development Studies, The Australian National University: Canberra.

McGinnis, MV, 1999 (ed) *Bioregionalism,* Routledge: London.

McMillan, DW and Chavis, DM, 1986, "Sense of community: a definition and a theory", *Journal of Community Psychology* 14 (January, 1986): 6-23.

Meinig, DW, 1979, "The beholding eye: ten versions of the same scene", in Meinig, DS (ed) *The Interpretation of Ordinary Landscapes: Geographical Essays,* Oxford University Press: New York.

M'Gonigle, RM, 1986, "The tribune and the tribe: toward a natural law of the market/legal state", *Ecology Law Quarterly* 13: 233-310.

Millar, J and Curtis, A, 1999, "The nature and role of farmer knowledge in temperate pasture management in the Murray-Darling Basin", *Rural Society* 9(1): 301-312

Millar, K, 1994, "Evaluating community involvement in revegetation: summary of results", School of Forestry, University of Melbourne: Melbourne, (Mimeo).

Mitchell, MK and Stapp, WB, 1995, "Watershed monitoring by students: education that improves water quality and the lives of people", in Chamala, S and Keith, K (eds), *Participative Approaches for Landcare: Perspectives, Policies and Programs,* Australian Academic Press: Brisbane.

Mitchell River Watershed Management Group, 1999, *Mitchell River Watershed Management Plan*, Mitchell River Watershed Management Group: Mareeba.

Morris, D, 1999, "Small is still beautiful", *Utne Reader* 93: 24-27.

MRWMG, see Mitchell River Watershed Management Group.

Newborough, JR (ed), 1996, "Sense of community", [special issue] *Journal of Community Psychology* 24(2).

Newby, H, Bell, C, Rose, D and Saunders, P, 1978, *Property, Paternalism and Power*, Hutchinson: London.

Nozick, M, 1992, *No Place like Home: Building Sustainable Communities*, Canadian Council on Social Development: Ottawa.

Olsen, M, 1965, *The Logic of Collective Action*, Harvard University Press: Cambridge, Massachusetts.

Ordubegian, L, 1993, "Talking with Ontario Breeding Bird Atlas volunteers", in Lerner, S (ed) *Environmental Stewardship: Studies in Active Earthkeeping*, Department of Geography, University of Waterloo: Waterloo, (Department of Geography Publication Series No, 39).

Osborne, D and Gaebler, T, 1992, *Reinventing Government: How the Entrepreneurial Spirit is Transforming the Public Sector*, Addison-Wesley Publishing Company: Reading.

Pakulski, J, Tranter, B and Crook, S, 1998, "The dynamics of environmental issues in Australia: concerns, clusters and carriers", *Australian Journal of Political Science* 33: 235-252.

Papadakis, E, 1996, *Environmental Politics and Institutional Change*, Cambridge University Press: Melbourne.

Pfueller, S, 1995, "Community and water quality initiatives in Gippsland", *Australian Journal of Soil and Water Conservation* 8(3): 26-32.

Phifer, BM, List, EF and Faulkner, B, 1989, "An overview of community development in America", in Christenson, JE and Robinson (Jr), JW (eds), *Community Development in Perspective*, Iowa State University Press: Ames.

Pinkerton, E, 1989, "Introduction: attaining better fisheries management through co-management - prospects, problems and propositions", in Pinkerton, E (ed) *Co-operative Management of Local Fisheries: New Directions for Improved Management and Community Development*, University of British Columbia Press: Vancouver.

Pitt, MW, and Yapp, TP, 1992 "Perceptions of land degradation and awareness of conservation programs in north-eastern New South Wales", in International Soil Conservation Organisation (ed), *7th ISCO Conference on Sustainable Land Management Practices*, International Soil Conservation Organisation: Sydney.

Powell, JM, 1976, *Environmental Management in Australia 1788-1914*, Oxford University Press: Melbourne.

Pretty, JN and Chambers, R, 1994, "Towards a learning paradigm: new professionalism and institutions for sustainable agriculture", in Scoones, I and Thompson, J (eds), 1994 *Beyond Farmer First: Rural People's Knowledge, Agricultural Research and Extension Practice*, Intermediate Technology Publications: London.

Quaid Projects Management, 1993, *A Welcome to the Mitchell River Watershed Management Group*, Quaid Projects Management: Mareeba, (Brochure).

Rahnema, M, 1992, "Participation", in Sachs, W (ed) *The Development Dictionary: A Guide to Knowledge as Power*, Zed Books: London.

Read, P, 1996, *Returning to Nothing: the Meaning of Lost Places*, Cambridge University Press: Melbourne.

Read, P, 2000 *Belonging: Australians, Place and Aboriginal Ownership*, Cambridge University Press: Oakleigh.

Relph, E, 1976, *Place and Placelessness*, Pion Limited: London.

Roberts, BR, 1990, *The Birth of Landcare: Collected Papers on Land Use*, University of Southern Queensland Press: Toowoomba.

Roberts, BR, 1992, "The significance and fundamental nature of landcare as a social movement", Land Use Study Centre, University of Southern Queensland: Towoomba, (Address to Tasmanian Landcare Awards Ceremony, Launceston, November 1992).

Robin, L, 1998, *Defending the Little Desert: The Rise of Ecological Consciousness in Australia*, Melbourne University Press: Carlton South.

Rose, DB, 1996, *Nourishing Terrains: Australian Aboriginal Views of Landscape and Wilderness*, Australian Heritage Commission: Canberra.

Rose, G, 1995, "Place and identity: a sense of place", in Massey, D and Jess, P (eds), *A Place in the World? Places, Cultures and Globalization*, The Open University, Oxford University Press: New York, (Explorations in Human Geography Series 4).

Rosenau, PM, 1992, *Post-modernism and the Social Sciences: Insights, Inroads and Intrusions*, Princeton University Press: Princeton, New Jersey.

Rush J and Associates, 1992, *A Review of the Efficiency of Landcare Facilitator Projects, Part 2: Landcare Groups, Part 3: Facilitators and Coordinators and Part 4: Final Review*, (A research report prepared for the Land Resources Division, Department of Primary industry and Energy, Canberra).

Sale, K, 1984, "Bioregionalism – a new way to treat the land", *The Ecologist* 14 (4): 167-173.

Sale, K, 1985, *Dwellers in the Land: The Bioregional Vision*, Sierra Club Books: San Francisco.

Sarason, SB, 1974, *The Psychological Sense of Community: Perspectives for Community Psychology*, Jossey-Bass: San Francisco.

Sarason, SB, 1993, "American psychology and the needs for transcendance and community", *American Journal of Community Psychology* 21: 185-202.

Schrecker, T, 1994, "Environmentalism and the politics of invisibility", *Alternatives* 20 (2): 32-37.

Scoones, I and Thompson, J, 1994, "Knowledge, power and agriculture: towards a theoretical understanding", in Scoones, I and Thompson, J (eds), *Beyond Farmer First: Rural People's Knowledge, Agricultural Research and Extension Practice,* Intermediate Technology Publication: London.

Scoones, I and Thompson, J (eds), 1994 *Beyond Farmer First: Rural People's Knowledge, Agricultural Research and Extension Practice,* Intermediate Technology Publications: London.

Shire of Serpentine Jarrahdale 1991, *Notes from the Local Rural Stragegy Workshop,* Shire of Serpentine Jarrahdale: Mundijong.

Star, J, 2000, "Time that orange", in Department of Environmental Protection (ed) *Bessie (ed) Brolly: A Celebration of Women and the Environment in Western Australia,* Department of Environmental Protection: Perth WA.

Steelman, TA and Carmin, J, 1998, "Common property, collective interests and community opposition to locally unwanted land uses", *Society and Natural Resources* 11: 485-504.

Summers, GF, 1986, "Rural community development", *Annual Review of Sociology Journal* 12: 347-371.

Syme, GJ, 1992, "When and where does participation count?" in Munro-Clark, M (ed) *Citizen Participation in Government,* Hale and Iremonger: Sydney.

Synnott, M, 1991, "Melbourne "88 plus two: a review of ICM in Australia", *Land and Water Research News* 8: 42-47.

Trzyna, TC, 1990, "Convening thinkers and doers: Sweden (ed) Dag Hammarskjöld Foundation", in Trzyna, TC and Gotelli, IM, *The Power of Convening: Collaborative Policy Forums for Sustainable Development,* California institute of Public Affairs: Sacramento, (Proceedings of an international workshop held at Claremont, California, 5-7 October 1989).

Tuan, YF, 1974, *Topophilia: a Study of Environmental Perception, Attitudes and Values,* Prentice Hall: New Jersey.

Turner, JC, 1991, *Social Influence,* Open University Press: Milton Keynes (Mapping Social Psychology Series).

United Nations, 1992, *Agenda 21: The United Nations Programme of Action from Rio,* United Nations: New York.

Vanclay, F, nd, *Stewardship and Conservation in Australian Farmers,* School of Australian Environmental Studies, Griffith University: Brisbane.

Vanclay, F, 1992, "The social context of farmers' adoption of environmentally sound farming practices", in Lawrence, G, Vanclay F, and Furze, B (eds), *Agriculture, Environment and Society,* MacMillan: South Melbourne.

Vanclay, F, 1994, "A crisis in agricultural extension", *Rural Society* 4 (1): 10-13.

Vanclay, F and Lawrence, G, 1995, *The Environmental Imperative: Ecosocial Concerns for Australian Agriculture,* Central Queensland University Press: Rockhampton.

Walker, KJ, 1987, "Methodologies for social aspects of environmental research", *Social Science Information* 26 (4): 759-782.

Ward, J, 1994, *Australian Community Development: Ideas, Skills and Values for the 90s*, Partnership Press and Community Quarterly: Melbourne, (Second edition).

Warner, PD, 1989, "Professional development roles", in Christenson, JE and Robinson (Jr), JW (eds), *Community Development in Perspective*, Iowa State University Press: Ames.

Waters, N, 2000, *The Community Planning Handbook: How People can Shape their Cities, Towns and Villages in any Part of the World*, Earthscan Publications Ltd: London.

Watson, HJ, Vallee, JM and Mulford, WR, 1981, *Structured Experiences and Group Development*, Curriculum Development Centre: Canberra.

West, R, 1992, *Making Groups Work*, NLP Consulting: Sydney.

White, T, 1992, "Land literacy", in Greening Australia (ed) *Catchments of Green Conference Proceedings Volume A*, Greening Australia: Canberra.

Wild, R, 1974, *Bradstow: A Study of Status, Class and Power in a Small Australian Town*, Angus and Robertson: Sydney.

Wilkinson, KP, 1989, "The future of community development", in Christenson, JE Robinson (Jr), JW (eds), *Community Development in Perspective*, Iowa State University Press: Ames.

Wynne, B, 1989, "Sheepfarming after Chernobyl", *Environment* 31: 10-23.

Wynne, B, 1992, "Misunderstood misunderstanding: social identities and public uptake of science", *Public Understanding of Science* 1: 281-304.

Wynne, B, 1995, "Public understanding of science", in Jasanoff, S, Markle, GE, Petersen, JL and Pinch, T (eds) *Handbook of Science and Technology Studies*, Sage Publications: Thousand Oaks, (published in cooperation with the Society for Social Studies of Science).

Wynne, B, 1996a, "May the sheep safely graze? A reflexive view of the expert-lay knowledge divide", in Lash, S, Szerszynski, B and Wynne, B (eds) *Risk, Environment and Modernity*, Sage Publications: London.

Wynne, B, 1996b, "SSK's identity parade: signing-up, off-and-on", *Social Studies of Science*, 26: 357-391.

Yearley, S, 1994, "Understanding science from the perspective of the sociology of scientific knowledge: an overview", *Public Understanding of Science* 3: 245-258.

Index